# Designing Healthy and Liveable Cities

In the last ten years, concepts such as urban health and liveability have become ever more present in urban planning studies. Many companies rank the most liveable city in the world or in a nation, and many indicators are used to try to measure factors which can report the health of a place by investigating it in different ways. While it is possible to understand why a place is liveable – due to the liveability and health concepts that are being more and more explored in urban studies, and the strong influence coming from other disciplines – it is difficult to design a place that is certain to be healthy and liveable.

Accordingly, aim of this book is, after the definition of the field of investigation concerning sustainable regeneration trough topics such as resilience, adaptation, health, and mixed connections, to illustrate the present-day approaches to the analysis and design of healthy places, and in particular the original Healthy Pl@ce Design method, flexible and repeatable in different contexts. The method aims to identify sustainable urban liveability and healthiness and the factors which make places liveable and healthy from users' points of view and identifying design interventions that can enhance or create both urban liveability and health. Emblematic case studies carried out in Europe, Canada and China – Bordeaux, Copenhagen, Hamburg, Madrid, Newcastle–Gateshead, Nice, Dublin, Vancouver and Wuhan – constitute the empirical part of the book, detailed with surveys, questionnaires, images and maps.

The theoretical framework – built on contemporary issues – and international case studies make this book both attractive and scientific, adding a new stone on the sustainable city construction and opening it to a particularly wide readership, including scholars, students, administrators and professionals.

**Marichela Sepe** is a researcher with the Institute for Studies on the Mediterranean of the Italian National Research Council in Naples, Italy. She joined the Department of Architecture of the University of Naples Federico II in 2003, where she serves as a contract professor of urban planning. Sepe is on the Steering Committee of the Italian Institute of Urban Planning and European Urban Research Association and is a member of the International Committee for Documentation and Conservation of Buildings, Sites, and Neighbourhoods of the Modern Movement and of the Urban Design Group.

## Routledge Studies in Urbanism and the City

**Overlooked Cities**
Power, Politics and Knowledge Beyond the Urban South
*Edited by Hanna A Ruszczyk, Erwin Nugraha, and Isolde de Villiers*

**The Walkable City**
*Jennie Middleton*

**Cities Without Capitalism**
*Edited By Hossein Sadri and Senem Zeybekoglu*

**Public Participation Process in Urban Planning**
Evaluation Approaches of Fairness and Effectiveness Criteria of Planning Advisory Committees
*By Kamal Uddin and Bhuiyan Monwar Alam*

**Hegemony, Security Infrastructures and the Politics of Crime**
Everyday Experiences in South Africa
*Gideon van Riet*

**Marketplaces**
Movements, Representations and Practices
*Edited by Ceren Sezer and Rianne van Melik*

**Designing Healthy and Liveable Cities**
Creating Sustainable Urban Regeneration
*Marichela Sepe*

**The Political Economy of Land**
Rent, Financialization and Resistance
*Edited by Mika Hyötyläinen and Robert Beauregard*

For more information about this series, please visit https://www.routledge.com/Routledge-Studies-in-Urbanism-and-t.he-City/book-series/RSUC

# Designing Healthy and Liveable Cities

## Creating Sustainable Urban Regeneration

**Marichela Sepe**

Routledge
Taylor & Francis Group
LONDON AND NEW YORK

First published 2023
by Routledge
4 Park Square, Milton Park, Abingdon, Oxon OX14 4RN

and by Routledge
605 Third Avenue, New York, NY 10158

*Routledge is an imprint of the Taylor & Francis Group, an informa business*

All figures, except those which are specifically indicated in the caption, are Author's images



*British Library Cataloguing-in-Publication Data*
A catalogue record for this book is available from the British Library

*Library of Congress Cataloging-in-Publication Data*
A catalog record has been requested for this book

ISBN: 978-0-367-56642-5 (hbk)
ISBN: 978-0-367-56644-9 (pbk)
ISBN: 978-1-003-09877-5 (ebk)

DOI: 10.4324/9781003098775

Typeset in Galliard
by SPi Technologies India Pvt Ltd (Straive)

To my family

# Contents

# Acknowledgement

This book was realized in the framework of the Research Project PRIN2020 #20209F3A37, Italian Ministry of University and Research (MUR).

The Chapter I of Part I was also realized in the framework of the research project LANDSUPPORT – H2020-RUR-2016-2017/H2020-RUR-2017-2.

Part III was also realized in the framework of COST Action CA18126: Writing Urban Places. New Narratives of the European City and COST Action CA18204: Dynamics of placemaking and digitization in Europe's cities.

The Wuhan case study was also carried out in the framework of the Research Project PLANNING, Urban Management and Heritage (PUMAH), FP7-PEOPLE-2011-IRSES-MC-IRSES and the CNR-CACH Italy–China Bilateral Agreement 'Cultural Itineraries and the One Belt One Road Initiative. Good practices and policies for conservation and tourism development'.

Finally, I would like to thank Faye Leering, Soujanya Ganesh and Wendy Jo Dymond at Routledge for their guidance and support in bringing this volume to publication and the anonymous reviewers who provided constructive advice to its improvement. Thanks also go to all colleagues, students and, more in general, people who are interested in healthy and liveable places, who in different ways gave me suggestions since the first ideation of this book arisen.

# Biography

**Marichela Sepe** (Naples, 1966) is a researcher at the ISMed – Institute for Studies on the Mediterranean of the Italian National Research Council of Naples. Since 2003, she is also with the Department of Architecture of the University of Naples Federico II, where she was a member (until 2017) of the Research Committee Doctorate in Urban Design and Planning and is a Contract Professor of Urban Planning. In 2013, she was a visiting professor at Peking University and held lectures in Peking, Wuhan and Xi'an Universities. Her research interests include sustainable urban design and planning, urban landscape, healthy cities, liveable public spaces, place identity, creative urban regeneration and cultural heritage enhancement and fruition. On these topics, she has published several national and international journal articles, conference papers, books and book chapters. Dr Sepe is Vice President of the Italian National Urban Planning Institute (INU) Campania section; was involved as a director in the 2013 INU Regional Urban Planning Exhibition, is on the national INU Steering Committee and on the EURA governing board and is member of the Docomomo and Urban Design Group. Between 2014 and 2018, she won the Ardito-Desio Award for her paper presented at Ipsapa 2014, 2016 and 2018 Conferences; the Runner-up Practitioner paper at the T-Forum 2015; and the Urban Planning Literature Award of the INU in 2014, 2015 and 2017. She has participated in several architecture and urban design competitions, in which she took two first places and one selection.

# Introduction

This book deals with the health of places and how to make them healthier, more liveable and sustainable from the point of view of the people who use them.

The urban conditions that prompted me to work on urban health topics (Sepe, 2020a) were not only deepening themes, such as climate change, different environmental issues and pandemics that are changing the world we live in (Sepe, 2021), but also understanding how to create places where the people who live there feel good, perceiving their healthiness and liveability.

In fact, it is not difficult to feel good in a place, but it is difficult to really identify the factors that make people feel good in that place, and it is even more difficult to understand how to design a place that is healthy and liveable.

Since the 19th century – in the period of the hygienists (Otgaar et al., 2011) – attention was paid to the influences of basic structures on public health. This topic then returned, after a pause, to the agenda of scholars, politicians and public administrators for various related issues such as the improvement of green spaces, public spaces and cycle and pedestrian networks as they are able to promote healthy behaviours in people towards themselves and the surrounding environment.

The health determinants (Otgaar et al., 2011) that can be identified include human biology, namely the factors internally related to the human body itself; the environment, namely the elements external to the human body and on which people have limited control, such as the social, natural and built environment; lifestyle, namely the healthy or unhealthy behaviour of people; and the health care facilities, namely the health care facilities and their quality and accessibility.

Since health is a concept studied in many disciplines, to illustrate it, I have limited the field of investigation to the observation of the factors and elements that make a place liveable and that can promote healthy behaviours in people, which, in turn, can contribute to the sustainable maintenance and improvement of spaces according to a virtuous circle (McHarg, 1969; Musterd and Kovács, 2013; Ruijsbroek et al., 2017; Sudjic, 2016; Whyte, 1980).

DOI: 10.4324/9781003098775-1

The book is divided into four parts. The first part is devoted to the definition of sustainable urban regeneration through a holistic approach that includes factors and elements such as resilience and adaptation (De Roo et al., 2020), mixed connections (Ratti, 2014), healthy urban places (Carmona, 2019a, b; Capolongo et al., 2018) and indicators, all introduced by the principles of the 2016 New Urban Agenda and the 17 Sustainable Development Goals. In this framework, urban health is considered a fundamental element for sustainable regeneration to be taken into consideration in not only all phases of transforming a place but also, and above all, afterwards. Indeed, the components that determine urban health are many; in this book, those concerning public spaces (Carmona et al., 2010) – which are also treated in the case studies – are especially considered. The healthiness issues are connected in different ways to place identity as well (Lynch, 1960; Cullen, 1961; Pendlebury and Porfyriou, 2017; Sepe, 2009a, 2013a, 2013b).

The second part is dedicated to emblematic contemporary methodological approaches to the analysis and design of healthy and liveable places and, in particular, 15-Minute (Moreno, 2020), flexible (Bergevoet and Van Tuijl, 2016), soft (Sim, 2019) and smart (Sassen and Kourtit, 2021) ones.

While the 15-Minute City is a city able to offer all its inhabitants everything they need to live, work and have fun to be reached on foot in no more than 15 minutes, the flexible one is based on tools for architectural and urban planning and design, which are able to allow changes in the course of implementation of those projects. The soft city is based on the idea that from the union of density and diversity a more liveable and healthier city can be obtained as the proximity of an environment can be translated into time. Finally, in the smart city, the whole range of technologies are at the service of the place both to improve its liveability and healthy and ensure its sustainability. These approaches to the study of places, albeit different, have several factors in common, including attention to integration, inclusion and diversity.

The third part is dedicated to the illustration of the original Healthy Pl@ce Design method (Sepe, 2020a) which aims to provide a repeatable investigation protocol to analyse and design healthy and liveable places. The protocol is made up of eight phases, five of analysis – in which the characteristics, equipment, organization and quality of the places and the way in which people experience and perceive them are obtained through surveys, analyses and questionnaires – and three of project, in which, after identifying factors and elements that can be improved, administering an on-site questionnaire and the realizing an analysis on social networks to understand people's opinions on project proposals, interventions are identified to obtain healthier and more liveable places.

To support the method, the characteristics and functions of a software program were created to visualize the analysis and project maps of healthy and liveable places in a faster and smarter way. Indicators have also been created to give numerical values to the healthiness of the analysed locations.

The fourth part presents the case studies in which Bordeaux, Hamburg, Madrid, Nice, Dublin, Vancouver and Wuhan are analysed with the Healthy

Pl@ce Design method and Copenhagen and Newcastle–Gateshead with secondary data. All the cases are emblematic of the cities in terms of liveability, urban health and sustainability.

The study area in Bordeaux is the left part of the Garonna waterfront which starts at Pont Saint Jean and ends at Pont Jacques Chaban Delmas. The case study was carried out in the Bordeaux area of urban regeneration that is based on the articulated planning of urban interventions and financial resources aimed at redefining the identity of the places, which has long been linked only to wine production, and revitalizing the economy of the city.

The Copenhagen case study concerns Superkilen park, created as part of Prequalified Competition launched by the Municipality of Copenhagen and the Realdania association, a Danish foundation which aims to finance or support projects that improve the aesthetics and liveability of the urban environment.

In Dublin, the area that is analysed is the Liffey waterfront between Grattan Bridge and Samuel Beckett Bridge and the area between Samuel Beckett Bridge and Grand Canal Quai–Grand Canal Bridge. The process of urban regeneration of this area started in the 1990s with the redesign of the pedestrian street along the River Liffey, the Samuel Beckett Bridge, the Bord Gáis Energy Theatre and its square.

The case study in Hamburg concerns the public spaces of HafenCity, a 155-hectare area involved in a wide urban regeneration operation which is started in 1998 after an idea competition. The aim of the regeneration operation was to allow the city centre to expand re-creating the maritime identity and, at the same time, integrate it with the port beyond the Elbe, which for decades has been increasingly relegated to the southern-most part of the town.

The Madrid case study is Madrid Rio, a 10-kilometre linear park completed in 2011 which is situated along the Manzanares River in the area that was before covered by the M30 highway. An important part of the area is the river, in course of renaturalization, and its ecological corridor. The objective of the regeneration includes the connection of all the river sections of the city enhancing both the beauty of the landscape – and its flora and fauna – and its social value in the urban context.

The Newcastle–Gateshead case study concerns the quayside, interested in a regeneration process aimed at urban change strongly linked to the local identity, paying attention to the participation and involvement of the population, the liveability of places, the long-term sustainability interventions and the creative economy. The general strategy and urban core plan which were adopted in March 2015 form a strategic planning framework for development in Newcastle and Gateshead up to 2030.

The Promenade du Paillon case study in Nice is related to an area which was realized in the framework of the operative programme PNRQAD (Programme National de Requalification des Quartiers Anciens Dégradés). The local administration decided to realize a regeneration of the Nice old area with an improvement of housing, public spaces and mobility with interventions which regard Place Massena (created in 1849 and renovated in 2007), Promenade du Paillon (inaugurated in 2013) and the Gare du Sud (inaugurated in 2019).

The Vancouver case study concern two new areas, namely Coal Harbour and False Creek. The regeneration process of these areas started in the 1990s in the framework of the Vancouver "Living First Strategy", focused on the improvement of the quality of life, walkability, public transport, local services and quality public spaces.

Finally, the new Hankou riverfront of Wuhan, whose development started after the last flood, in 1999, with both the functions to create flood control and a quality urban design. The waterfront area of analysis is about 4 kilometres and has a linear plan to be constituted for both green and built areas. The greater particularity is that it borders five concession areas that create a variety of architectural styles.

The last part of the book is devoted to the conclusion, that concerns the Principles of Urban Healthy and Liveability Design, derived from the case studies carried out with the Healthy Pl@ce Design method. These principles are a sort of checklist to take into account both to analyse and design places which have health and liveability as their main factors to achieve sustainability.

# Part I

# Sustainable urban regeneration

# 1 Holistic approach

Urban regeneration consists of an integrated approach between vision and action for the resolution of various problems related to disadvantaged urban areas in order to improve their socio-economic, physical and environmental conditions with actions such as the requalification, recovery and conservation of heritage (Zheng et al., 2017).

To the term *regeneration* must be added the term *sustainable* (Nijkamp and Perrels, 1994) understood in its triple meaning (Sepe, 2009b, 2013b, 2014a, 2018; Sepe et al., 2016), although as Evans and Jones (2008) affirm, it can create ambiguity on what weight to give to the environmental, social and economic component, determining that greater emphasis can be placed in the regeneration processes on one element instead of another depending on the developers' goals (Astleithner et al., 2004; Davies, 2002).

Indeed, in the most recent studies for this purpose, additional key elements of sustainability in relation to the planning system have been identified, namely cohesion and social inclusion, protection and enhancement of the natural environment, the prudent use of natural resources, sustainable economic growth and integration of sustainable development into development plans (ODPM, 2004). This highlights the importance of integrating the components and not just ensuring their presence (Forrester and Snell, 2007).

In the chapters of Part I, some elements and factors considered to be important for determining sustainable urban regeneration are illustrated, including resilience and adaptation to crises and changes, the healthiness of places and new connections and indicators, all following the holistic approach detailed here that frames the general points of view that led to the identification of these elements and how these were treated in the book.

Each transformation operation may have to undergo changes due to sudden crises of different types; it is important to be able to guarantee the resilience of places and their adaptation which require all the involved components to be flexible at all levels.

Urban health is a fundamental element for sustainable regeneration and must be taken into consideration in not only all phases of transforming a

DOI: 10.4324/9781003098775-3

place but also, and above all, afterwards. The components that determine urban health are many; in this book, those concerning public spaces – also treated in the case studies – are especially considered.

The kinds of connections between people and place are increasingly growing. The social infrastructure in urban regeneration areas is an important element in sustainable development and must have equal access for residents. It is also important to remember that stakeholders do not have equal rights and powers in the transformation process. Stakeholders in urban regeneration projects must include not only public administrators at different levels and stakeholders in the private sector but also citizens and, in particular, those who live close to urban renewal projects.

A dynamic relationship must be created between different actors who share objectives agreed together and which must be pursued through the understanding of mutual needs and benefits (Brinkerhoff, 2002). Similarly, the creation of green and blue infrastructure is fundamental to guarantee ecological sustainability and economic regeneration and create a suitable balance with the grey ones. Cultural networks are another important element in sustainable regeneration as they have the potential to improve social cohesion and the city's brand. Accordingly, planning attentive to perceptions of one's heritage can improve interaction with this and the surrounding urban environment, with a positive effect on the health and liveability of the place. Consequently, the integration of all the infrastructures will be the key factor to ensure more sustainable connections.

The monitoring of the regeneration process throughout its path is also important to obtain more sustainable results and improve existing programs by modifying the proposed solutions with others that are more adequate than the actual needs. The evaluation of sustainable urban renewal is useful for making stakeholders understand which are the most suitable strategies to effectively obtain sustainable regeneration. In this regard, the indicators can be a useful reference for assessing the usefulness of regeneration programs and their effects on urban health and liveability.

Indeed, the holistic approach to the design of places is an approach that considers all the components that determine their implementation in terms of sustainability, healthiness and liveability, as well as all those involved in a logic of integration and equity. The holistic approach (Lehmann, 2010; Xu, 2011) can be read in various Charters, agendas, principles and international reports which in recent decades have formed the framework within which local agendas have been built. Some principles of the New Urban Agenda and of the 17 Sustainable Development Goals (SDGs) which constitute an important reference for the contents of this book are reiterated in the following.

Many principles that are contained in the New Urban Agenda concern topics related to public spaces, liveability, healthiness and accessibility. It was adopted during the UN-Habitat III Conference which was held in Quito, Ecuador, in October 2016. It represents "a shared vision for a better and more sustainable future" (UN-Habitat, 2016).

Accordingly, in the following, a selection of these principles is reported.

> 36. We commit ourselves to promoting appropriate measures in cities and human settlements that facilitate access for persons with disabilities, on an equal basis with others, to the physical environment of cities, in particular to public spaces, public transport, housing, education and health facilities, public information and communication (including information and communications technologies and systems) and other facilities and services open or provided to the public, in both urban and rural areas.

This principle sets out a very important element for the health of places, that is facilitating access for people with disabilities in public spaces and built structures but also in the sphere of communication and information, allowing inclusive access to tangible and intangible resources of a territory.

> 37. We commit ourselves to promoting safe, inclusive, accessible, green and quality public spaces, including streets, sidewalks and cycling lanes, squares, waterfront areas, gardens and parks, that are multifunctional areas for social interaction and inclusion, human health and well-being, economic exchange and cultural expression and dialogue among a wide diversity of people and cultures, and that are designed and managed to ensure human development and build peaceful, inclusive and participatory societies, as well as to promote living together, connectivity and social inclusion.

The important points that are expressed in this principle concern the promotion of public spaces of all types that have different functions – an important aspect even in times of pandemics in which it is important to have different uses to be carried out outdoors – and that are designed and, above all, managed to promote a dialogue between all. Public space, if it is of quality, is also able to promote not only health and well-being in those who live it but also economic and cultural exchange.

> 50. We commit ourselves to encouraging urban-rural interactions and connectivity by strengthening sustainable transport and mobility, and technology and communications networks and infrastructure, underpinned by planning instruments based on an integrated urban and territorial approach, in order to maximize the potential of these sectors for enhanced productivity, social, economic and territorial cohesion, as well as safety and environmental sustainability. This should include connectivity between cities and their surroundings, peri-urban and rural areas, as well as greater land-sea connections, where appropriate.

Both physical and virtual infrastructures, as illustrated in one section of this book, are important elements for connecting places, especially if they are understood in terms of sustainability and if they relate to all urban, rural and marine areas.

> 67. We commit ourselves to promoting the creation and maintenance of well-connected and well distributed networks of open, multipurpose, safe, inclusive, accessible, green and quality public spaces, to improving the resilience of cities to disasters and climate change, including floods, drought risks and heat waves, to improving food security and nutrition, physical and mental health, and household and ambient air quality, to reducing noise and promoting attractive and liveable cities, human settlements and urban landscapes and to prioritizing the conservation of endemic species.

In this principle, the issue of resilience to environmental disasters is associated with that of creating networks of public spaces, understood as places that can help improve resilience along with other aspects such as air quality and the liveability of urban landscapes.

> 97. We will promote planned urban extensions and infill, prioritizing renewal, regeneration and retrofitting of urban areas, as appropriate, including the upgrading of slums and informal settlements, providing high-quality buildings and public spaces, promoting integrated and participatory approaches involving all relevant stakeholders and inhabitants and avoiding spatial and socioeconomic segregation and gentrification, while preserving cultural heritage and preventing and containing urban sprawl.

In this principle, the theme of sustainable regeneration includes, in addition to quality public spaces, the promotion of integrated participation operations and the safeguarding of cultural heritage. The important aspect that can be seen in this, as well as in the other principles, is that each topic treated is linked with several others precisely from the perspective of the holistic approach mentioned earlier.

> 118. We will encourage national, subnational and local governments to develop and expand financing instruments, enabling them to improve their transport and mobility infrastructure and systems, such as mass rapid-transit systems, integrated transport systems, air and rail systems, and safe, sufficient and adequate pedestrian and cycling infrastructure and technology-based innovations in transport and transit systems to reduce congestion and pollution while improving efficiency, connectivity, accessibility, health and quality of life.

This principle highlights the improvement of all transport and mobility infrastructures in the logic of integration on the one hand but also of safety and healthiness on the other. Reducing air pollution due to transport systems also has a positive effect on slow mobility systems and therefore on the quality of life.

Before the III Un-Habitat Conference, in 2015 the General Assembly of United Nations Member States adopted the 2030 Agenda for Sustainable Development and 17 SDGs it provides (UN, 2015). The Goals contain strategies to improve health and education, reduce inequality and incentivize economic growth. These include (1) ending poverty in all its forms everywhere; (2) ending hunger, achieving food security and improved nutrition and promoting sustainable agriculture; (3) ensuring healthy lives and promoting well-being for all, at all ages; (4) ensuring inclusive and equitable quality education and promoting lifelong learning opportunities for all; (5) achieving gender equality and empowering all women and girls; (6) ensuring the availability and sustainable management of water and sanitation for all; (7) ensuring access to affordable, reliable, sustainable and modern energy for all; (8) promoting sustained, inclusive and sustainable economic growth, as well as full and productive employment and decent work for all; (9) building resilient infrastructure, promoting inclusive and sustainable industrialization and fostering innovation; (10) reducing inequality within and among countries; (11) making cities and human settlements inclusive, safe, resilient and sustainable; (12) ensuring sustainable consumption and production patterns; (13) taking urgent action to combat climate change and its impacts; (14) conserving and sustainably use the oceans, seas and marine resources for sustainable development; (15) protecting, restoring and promoting sustainable use of terrestrial ecosystems, managing forests, combating desertification and biodiversity loss and halting and reversing land degradation; (16) promoting peaceful and inclusive societies for sustainable development, as well as providing access to justice for all and building effective, accountable and inclusive institutions; and (17) strengthening the means of implementation and revitalizing the global partnership for sustainable development.

These 2030 Agenda goals present a holistic approach as the principles concerning the 2016 New Urban Agenda. Topics such as resilience, equity, health, safety, sustainable economic growth, resilient infrastructures and nutrition improvement are some of those in common between the two Agendas and constitute, as mentioned before, the wider framework of the topis of this book as well.

# 2 Resilience and adaptation

*Resilience* and *adaptation* are two closely related and interconnected terms and have several points in common, as deduced in the following, so these are illustrated here together.

The term *resilience* has ancient origins, dating back to at least the first century B.C. and its first use was found in Lucretius's "On the Nature of Things" in reference to Nature's action comparable to the "bounce" of an image in a mirror (Pizzo, 2015). Since then, the term has been used in a variety of contexts, but it is in the context of Holling's (1973) studies in relation to ecological systems that the concept of resilience changes and refers to the ability of systems to absorb change and maintain the same relationships between populations or state variables.

In subsequent years, the term has been used in many other disciplines in part because of its adaptability to complexity theories, including economics, ecology, political science, cognitive science, new technologies and land-use planning. Accordingly, Zolli and Healey's (2012) definition of resilience understood as "the ability of a system, firm, or person to maintain its core purpose and integrity in the face of dramatically changed circumstances" integrates ecology and sociology and gives an insight into the multidisciplinary nature of the issue.

Resilience often considers events that cause a crisis in a system to be the same, even though these have different characteristics, and does not differentiate sudden events from undesirable ones, leading to some ambiguities. Among them, for example, since there is no definition of how to pursue it, the degree of resilience of communities after a sudden event may not be the same, and even within the same community, different behaviours may occur.

Considering an evolutionary approach (Bohland et al., 2019; Davoudi et al., 2012; Russo, 2018), there is no single equilibrium in ecology, but there are multiple equilibria; however, the issue to focus on is the type of equilibrium to refer to rather than the causes that could alter it.

A crisis of social or cultural type must be faced with different approaches than one due to a catastrophe of environmental type, and moreover, even the same type of crisis can have different durations of effects and variable response times. The characteristic of multidimensionality of resilience makes

DOI: 10.4324/9781003098775-4

the concept of resilience flexible, on one hand, and elusive in its entirety and complexity, on the other, requiring continuous updates and insights. The current studies are focusing on environmental risks and, more recently, on those due to epidemics as in the case of COVID-19, allowing to highlight more clearly the inherent problems (Vale and Campanella, 2005).

It is also true, as Vale (2014) writes, that resilience is used in different fields in similar ways. Management analysts use resilience to measure a company's ability to recover from a disruption of a key element and recover as usual, just as economists measure it in relation to a place's ability to recover after the loss of a particular industry, ecologists are concerned with how a system can be restored to its previous equilibrium after a sudden environmental event (Banica et al., 2020; Borsekova and Nijkamp, 2019), psychologists use resilience to describe an individual's ability to withstand trauma and continue to function well. Information technology (IT) professionals measure the resilience of a communication network by relating how effectively a communication network copes with an outage exemplified or a massive power outage.

Within this framework, ecologists differ on the idea that there is a limit, that is once systems exceed the limit they collapse into a new state where there has been an extinction of species and a different set of processes are created. The concept of resilience in this direction becomes closer to that of non-equilibrium and can also be applied to interpretations of cities (Pickett et al., 2004).

Applying the concept of resilience to socio-environmental systems such as cities means anticipating crises and strengthening cities with proactive solutions that can at the same time enhance both public and private places.

However, preventive resilience requires choices that require upfront spending and investment and decisions about which people and places are at risk and which should be protected, regardless of whether the hazards are human-made, natural or a combination of them. A holistic view of preventive resilience should include consideration of the needs of all stakeholders and the different types of places potentially affected.

Resilience is also a concept that in addition to environmental quality is combined with the definition of safety in which a resilient place is a place where people live safely within improved places.

Indeed, cities are not uniform landscapes of people that are randomly distributed but organized places that produce socio-economic differences; therefore, it is not easy to describe a city as resilient in its entirety. For example, the COVID-19 pandemic brought problems of safety, fragility and differences in resilience not only between the cities affected but also between areas within them.

Certain factors and their components are important in determining resilience, including vulnerability analysis, which has uncertainty, informality, demographics and urban distribution of vulnerability as components; uncertainty-oriented planning, which has adaptation, planning and sustainable form as components; urban governance, which has equity, integration and ecological economics as components; and prevention, which has mitigation, restructuring and alternative energy application as components.

Demographic variables influence the ability of individuals to cope with environmental risks and uncertainty; socio-economically weak communities are, in fact, more prone to suffer negative effects both in terms of material and immaterial damages (Ojerio et al., 2011).

Informal urban spaces are more vulnerable because in many cases, there is a lack of infrastructures and services and the presence of a large population with socioeconomic hardship, which can result in a wide variety of risks and less resilience.

Uncertainty is another important component in the study of resilience because a lack of knowledge about future crises of an environmental nature and beyond makes people and places particularly vulnerable. To mitigate the impacts due to uncertainty of future events, policy and urban risk planning and management must be taken into account in advance.

Integrating uncertainties within the planning process and improving collaboration between different institutions and organizations, both public and private at all levels, are key components of resilience. Similarly, social diversity must be considered in order to reduce it through a more balanced distribution of resilience resources.

A green economy can play an important role in achieving urban resilience that can be translated into investments directed toward energy efficiency and markets that stimulate eco-friendly consumption (David, 2006). Prevention is assuming an increasingly important role in urban resilience especially if thought in terms of risk mitigation – which can be translated into a reduction of not only greenhouse gas emissions but also social inequalities and imbalances in terms of urban decay – and restructuring – that is capacity and flexibility to concretely face ecological, cultural, economic challenges with specific strategies and policies for transformations in cities.

Resilient planning should therefore be oriented towards uncertainty and go beyond traditional approaches, preparing cities for possible changes. Adaptation in this sense is key to limiting damage from climate change and, more broadly, from economic, social and health crises. Planning should include a wider range of conditions and, in particular, develop *ex-ante* and *ex-post* analyses to adapt appropriately to sudden situations.

Another component of uncertainty is the sustainable urban form, which relates to many aspects of resilience: liveability, urban health, change, climate and multiculturalism are just some of the elements that influence contemporary public policy and must be taken into account for a good city form (Lynch, 1984). Urban compactness and contiguity, high-density planning, sustainable transportation and equitable access, mixed land uses, diversity of housing and built form, passive solar design, greening in particular cities and renewal and use with the rehabilitation and refunctionalization of brownfield sites constitute key criteria for assessing the sustainability of urban form (Jabareen, 2006, 2013).

In this regard, as Desouza and Flanery (2013) observe, the social sphere is composed of three types of elements, namely people, institutions - understood as the set of individuals converging towards common goals -, and activities - meant as the tasks that people and institutions design, carry out and use. In this framework, people play the most important role because it is around them that the other two components are built. Moreover, the physical and social spheres often overlap, and many activities within a city occur while encountering these two spheres. In public spaces for example, interactions between different people, activities (walking) institutions (services), processes (permits of various kinds) and resources (artwork and artefacts of various kinds) can be observed. Each of these components interacts in intentional and unintentional ways that are important for identifying the creation of resilient and adaptive places (Desouza and Flanery, 2013).

Spaces with a good degree of resilience are spaces that are able to adapt to change. It is also true that ***adaptivity***, like resilience, is a term that lacks an unambiguous interpretation (De Roo and Porter, 2007). Adaptive planning, adaptive governance, adaptive management, adapting institutions and adapting the city, its public spaces and its buildings are some meanings used to define this concept, giving different meanings to the concept of adaptivity.

To clarify its limits in the planning domain, it is possible to refer to this domain (De Roo et al., 2020) as including the material dimension and both the organizational and institutional dimensions. The material dimension is about people, the environment and the land, while the organizational and institutional dimensions are about the domains in which planning and activities act to link institutions with the material dimension. Adaptive planning can thus involve that the people, places or situations that are the objects of planning have dynamic behaviour and can exhibit that behaviour in the future or that the processes of designing and implementing interventions are capable of being adaptive.

Traditional approaches to planning do not consider adaptive behaviour as an initial point in the planning phase. These are situations that require interventions that are decided on later through planning action, such as for example building a road to improve deteriorating traffic conditions or building a residential neighbourhood in response to an increase in population.

These cases could have been resolved even early in the planning process; instead, traditional planning deals with planning based on current realities not on unforeseen changes.

The uncertainty and unpredictability of an event and its subsequent, often unknown, development are difficult elements to manage. In fact, the system as a whole cannot be understood only by observing the parts that compose it, which must instead be analysed together with their context, noting the reciprocal relationships and the ways in which the system reaches the best possible configuration.

Indeed, traditional approaches to spatial planning often do not take adaptive behaviour as a starting point, believing that urban interventions can be decided on the basis of facts and estimates that are available at the time of decision-making. If reality did not change by remaining similar to that considered during the decision-making process or otherwise predictable, there would be no reason not to continue with this form of planning. This form of planning is based on a static perspective, which assumes a transformation according to predictable patterns, whose starting points are: "actual, eliminating the anomaly responsible for the disturbed order in the here and now"; desired, that is ideal and evidence of context; and potential, that is the tools to achieve a predefined end on which there is consensus.

This static perspective is also used in contemporary planning, where deliberate actions will lead to the results of the decision intended. However, the assumption of a static world can lead to strategies that may be obsolete at the time of the decision, resulting in much divergence between expected and actual effects.

Resilience and adaptation are to be considered fundamental in dynamic planning, where the uncertain and the sudden become components of the planning process with the same weight as the others and where urban planning tools are renewed or modified in order to contribute to the management of any crisis in an appropriate way. While it is true that unforeseen events – by definition – occur suddenly, it is also true that adaptive and dynamic planning can better support the creation of a new equilibrium.

# 3 Healthy urban places

The relationship between health and urban design is complex because of the multiple elements which play different roles in the city system.

Urban places are generally made up of buildings, roads, open spaces and landscapes and the people who live in them, and can therefore be defined as socio-physical constructs (Carmona, 2019a).

As several international studies on places (Speck, 2012) report, places influence the opportunities available to people, happiness (Montgomery, 2013) and sense of belonging and are important for cultural and social activities (Barton, 2016; Bulkley & Henig, 2010; Corburn, 2009; Cushing and Miller, 2020; Dannenberg et al., 2011; Hall, 2014; London, 2020; Madanipour, 2011).

Furthermore, the location influences housing conditions and the real estate market, as well as physical and mental well-being and health (Adams and Tiesdell, 2013; Montgomery, 1998).

To explain the value of a place from a health, social, economic and environmental point of view, Carmona (2014, 2019a, 2019b) uses the term positional value, defined as "a measure of the value of something". In particular, England identifies six different types of values (CABE, 2006), which include the exchange value of parts of the built environment; the use value relating to the impact of the built environment on the activities that are carried out; the image value, which concerns the identity of the built environment projects; the social value and how much the built environment supports or is a threat to social relations; the environmental value and its resources; and the cultural value of the built environment.

In this way, the value of the place expands and refers to a complex and interconnected system of positive aspects to be connected to the interventions in the built environment and to all those who have different interests in the place and in its quality.

The definitions of the quality of the place indicated in the scientific literature are often related to terms such as *liveability*, *quality of the built environment*, *sustainability* and *urban design*, according to the different practical and political traditions. Several other terms can be connected to the concept

DOI: 10.4324/9781003098775-5

of the quality of place. Carmona et al. (2010) for example define 12 local environmental qualities: *clean and tidy*, *functional*, *distinctive*, *safe and secure*, *robust*, *green and unpolluted*, *accessible*, *attractive*, *comfortable*, *inclusive*, *vital and viable*, and *fulfilling*, which, in turn, are connected to a broad set of other terms and issues.

The World Health Organization recognizes that elements of the built environment such as air quality, green and blue areas, urban waste management and water supply have direct and indirect impacts on the protection from climate change as well as on the health of people (Capolongo et al., 2018).

Adequate living conditions for citizens require healthy environments that promote healthy lifestyles. The urbanization that is affecting the most densely built-up areas will lead 60% of the world population to live (Talukder, et al., 2015) by 2050 in medium-large cities, which, on one hand, meet the needs of the growing population and, on the other, create problems from the point of view of pollution, the heat island effect and the loss of urban green spaces. The result is environmental disasters of various magnitudes that have environmental, economic, social and human life and health implications, which could force people to migrate to other urban areas with less adverse weather phenomena and better social and health conditions.

Indeed, people's health is connected to urban policies that establish the link between form and function of the urban context and health (Fehr and Capolongo, 2016), with opportunities and risks to be carefully evaluated. Globalization and urbanization, combined with an ageing population, can expose people to risks of unhealthy behaviours, including reduced physical activity and unhealthy food, which can, in turn, cause chronic diseases. Proper planning of the urban environment and public spaces is an important factor in this sense for public health.

Disease prevention can therefore also be implemented through the adequate promotion of the health of places (Capolongo et al., 2018) which can be summarized in some key actions that include encouraging the implementation of interventions that facilitate healthy behaviour of people; improving living conditions in urban contexts; building cities that are attentive to older adults, accessible and inclusive; creating resilient urban areas; creating new economies and jobs through urban interventions; eliminating social inequalities; improving awareness of the beneficial effects that public health can have in cities; creating participatory public governance; introducing performance systems to measure the promotion of healthy living attitudes in cities; and improving the sharing information. These key actions have been related to objectives and related sub-objectives, which include environmental and social sustainability of urban areas, paying attention to soil consumption, the management of solid urban waste, and urban environmental planning; adaptation to climate change, with attention to the management of adverse weather events, urban resilience and housing emergencies due to socio-economic and environmental changes; and responses to the needs of the population, with reference to new forms of living, the perception of urban comfort and protection, urban greenery and infrastructure. For each of the sub-objectives, different strategies that have many points in common and interdependence can be connected for achieving the sub-objective.

The mitigation of climate change can be implemented with different types of strategies; among these, some that can be particularly effective concern the improvement of the communication tools of the alarms relating to adverse weather events in order to reduce mortality, the conservation of forests and green plants and the improvement of access to food in case of an environmental emergency. The latter strategy can be implemented by creating plans for food resilience in vulnerable areas and enhancing healthy environments.

Indeed, it is true that urban sprawl is also a factor in climate change and in atmospheric and noise pollution that can cause negative effects on people's quality of life. The strategies that can be implemented in this regard to reduce unplanned urban development include environmental monitoring, evaluation of the sprawl index in urban plans and in relation to socio-economic variables and monitoring of uncontrolled urban expansion, useful for the design of public green spaces for various health promotion functions, including physical activity (D'Alessandro et al., 2015).

Housing policies are another fundamental factor in ensuring the health of citizens, paying attention not only to the access to decent housing but also to its sustainable context from a social, cultural and environmental point of view. The strategies that can be adopted in this sense include the optimization of the quality/price ratio, the promotion of new forms of housing that offers the possibility to the inhabitants to participate in the creation of a sustainable community and that has the presence of particular services for older adults, children and non-self-sufficient people (Capolongo et al., 2011).

It is also important that there is an adequate socio-assistance network on an urban scale that connects public transport, neighbourhood planning and the usability of open spaces, as well as a vision of public health that incorporates a perspective of equity through the elimination of linguistic, cultural and social barriers.

Recent studies on these topics (McCay, 2017) affirm that firstly it is important to identify reasons why people who live in cities are at greater risk of having health problems because of different factors, including disparities; crowding, noise and pollution, which can produce stress and encourage people to avoid social relationships that are important for mental well-being; the low presence of green areas; and a possible reduction of leisure time, security and privacy.

Accordingly, green areas play a very important role in the environmental quality of cities and for people as they can affect air quality, mental health (stress and physical activity levels), social cohesion (Capolongo et al., 2018). The strategies for planning and managing green areas include planning and managing green areas correctly also from a network perspective, choosing essences capable of having an impact on the habitat (ability to absorb $CO_2$); increasing the perception of a space's safety with adequate lighting and services; improving accessibility, reachability and usability with diversified pedestrian and cycle paths for different types of users; and using flooring, quality urban equipment and involvement of the local community in the design of green areas.

The Mind the GAPS framework, created by McCay (2017), identifies four topics to make tangible the urban planner help, even though no one city still embodies all of them. The first is, as affirmed in different theories, access to urban green places, which can have many good factors, including encouraging exercise and social interaction. Wilson (1984) affirms that the good effect of green space stands in the fact that humans have a biological need satisfied by the contact with other species; Ulrich et al. (1991) state that the good effect is due to the contemporaneity of experience with aesthetics of nature and distance from everyday problems; the Kaplan (Rachel and Stephen, 1998) theory proposes that natural sites take the attention of people more than the non-natural because the latter need contemporary concentration on many elements.

These theories can help with understanding the positive effect of green space on mental health but are not enough to assume that the design of green space in a site can assure improvement in mental health.

There are other factors that can negatively influence the benefits of green spaces, such as inaccessibility, bad management and a threatening feeling that can discourage its use. Green spaces should be designed to be welcoming to different kinds of people, not monopolized by certain groups. Indeed, in some cases, green spaces can encourage behaviour which is anti-social and discourage their use because people feel unsafe in those spaces.

Successful green spaces should be walkable and have both suitable physical and visual accessibility, factors which improve people's well-being. Accordingly, a small presence of green in a small space, the streets or the workplace is important as well.

Another important factor is to create different options of transport such as paths for pedestrians and bikes, which help reduce sedentary habits.

Furthermore, it is important to design public spaces with flexible use which are capable of creating a sense of belonging and community, such as providing street furniture for both resting and chatting and other elements which contribute to social activities and the general perception of well-being.

The perception of safety contributes to a better quality of life and public spaces which are capable to contribute to this feeling are successful. Some features which can help this feeling include – as aforementioned – suitable lighting and clear landmarks.

As Lucy Saunders (2017) affirms people in the last decades are living longer than ever and so it is necessary to have healthier places to allow healthier lives for all. This is because daily activities, such as work, travel and leisure, are directly or indirectly connected to the place where these occur, requiring suitable spaces. Elements such as air pollution, noise, road dangers and social connectedness can influence the health of people in different ways. The question is to understand what of these factors influence health in a wider way and how to address them. Sauders suggests addressing all of them at once and implementing good practice in urban design at the largest possible scale.

Places which are good for people are often healthy as well. In the public realm, this often means diminishing motorized transport, which is used by people who could instead walk or cycle. Motorized transport impacts in non-positive ways air pollution and physical activity, transforming public spaces into noisy and unwelcoming areas and increasing the problems deriving from poor access and dangerous roads, which are particularly important for the most disadvantaged people. Putting people at the centre of the urban design and decision-making potentially reduces inequalities. Accordingly, Healthy Streets, as Lucy Sauders suggests, identifies 10 indicators which indicate how design urban spaces and transport policies put people first.

The most two important indicators – pedestrians from all walks of life and people choosing to walk and cycle – have the aim of creating places in which all people can participate in public life and both healthy and friendly environment design are considered a priority.

The other eight indicators – people feeling relaxed, easy-to-cross streets, clean air, not too noisy, places to stop and rest, people feeling safe, things to see and do, shade and shelter – show what is necessary to do to create inclusive, appealing and healthy places. Places which do not have these characteristics discourage people in using them. All professionals – transport professionals, landscape architects, community groups, artists, planners and developers – are involved in the realization of Healthy Streets, and all the indicators are necessary for creating them.

A big challenge is to meet the demands of different stakeholders in order to put into practice the principles, all often in limited spaces. Streets can be contested spaces, and every street requires a different approach according to social, physical, political or financial questions, which can add weight to the general design but not always in the same way. Important, as declared by Sauders, is that the general objective is to create public spaces which work better for people and are healthy and liveable for all.

While recognizing the importance of liveable public space for keeping people in good health, it is difficult to demonstrate that good urban design is capable to reduce - for example - heart disease. The increasing number of people who spend time, walk and cycle on the streets can be shown easily and is a first measure for demonstrating the success of the environment from a health point of view. There exist important factors which create good interactions between places and behaviours, including the promotion of daily activity, such as walking or cycling, which helps people have a healthy weight; the facilitation of easy access to healthy food; and the promotion of connected neighbourhoods and economic development (Figure 3.1).

*Figure 3.1* Paris, Parc Rives de Seine, riverside promenade

Richard Crappsley's (2017) approach to healthy design is mainly focused on streets as well and includes reducing on-street parking and reallocating road space for vehicles while adding space for walking and cycling; improving both the quality of footway zones and furniture and making it easier for people of all ages cross the streets; making bicycle infrastructure safer and more comfortable as well as introducing suitable separation with busy roads; slowing traffic by using different kinds of visual elements and tight corners; improving environmental elements, such as trees, sustainable drainage, shades streets and reduction of traffic noise; adding on-streets activities with both formal and informal activities which encourage people to use the space for social interaction or spend more time there; making safer the streets dissuading crimes and sense of insecurity, improving suitable lighting, natural surveillance; making streets welcoming to all; and improving accessibility.

The different approaches which were illustrated in this section have many factors in common. What seems clear is that the issues inherent in healthy places are many, with a lot of interrelationships and should be treated with strategies at different urban scales to promote healthier places and behaviours. The COVID-19 pandemic has underlined even more the relationships between elements on a global scale and factors on a local scale, between the quality of places and the quality of life of people, highlighting the importance of open public spaces both as spaces in themselves and as spaces of relationships between people and built environments.

# 4 Mixed connections

The infrastructural corridors built today by land, sea and air connect cities, towns and territories, often creating not only new geographies and new neighbourhoods but also unexpected distances.

The system of networks and infrastructures that a city presents is complex as it intersects not only the worlds created by new technologies but also those of social and cultural relations.

The changes in the climate have begun, first slowly and then more and more rapidly, to make their effects felt. Natural disasters have separated territories, people and economies but have also given stimulus for an evolution in the urban planning disciplines that have to necessarily update the toolbox to include tools for adapting to different types of crises, not only environmental.

Increasingly sophisticated grey infrastructures protect against river and coastal flooding and constitute drainage systems for stormwater management (Depietri and McPhearson, 2017).

The blue nets constitute the visible and invisible texture of many territories: waterways, riverbeds, rivers, seas, lakes and streams change the perspectives of places with their presence, constituting an opportunity for regeneration and enhancement. These connect, reconnect, revitalize and defend.

Green networks, with their shapes, landscapes and products of nature, create chains of territories and mosaics of places, giving life, as do blue networks, to a continuity of meaning even beyond institutional borders.

Urban parks for example are an essential element of the network of green – and blue – spaces and are also vital for human social (Croucher et al., 2007; de Vries et al., 2003; de Vries et al., 2013; Dines, et al., 2006), economic and environmental well-being (Benedict and McMahon, 2006; Gallo et al., 2015; Mitchell and Popham, 2008).

Ecological corridors and infrastructural corridors, green networks and blue networks, urban planning and adaptive projects seeking balance find new resilience and urban quality. In this perspective, resilience should not be interpreted only in physical terms (Brownson et al., 2001) but also in social

DOI: 10.4324/9781003098775-6

and economic ones, capable of determining conditions of social cohesion and stimulating processes of cultural awareness and self-organization (Weijers et al., 2016).

It is true that there is no clear distinction between the different types of networks. The infrastructural networks constitute the physical subsystem of the city, the venous system that distributes the flows for the survival of the socio-anthropic component and the urban system. These arise from the project design and must ensure a satisfactory and efficient level of service to the urban community. These nets are material, not very flexible and highly vulnerable, due to their intrinsic rigidity. In many cases, the presence of the infrastructural structure marks the limit of the city.

If, on one hand, the effects of physical infrastructures are in some way predictable or at least attributable to comparable phenomena, it is different for virtual, social and cultural networks that open to increasingly large and uncontrollable ecosystems, with intersections and overlaps that mimic the real relationships but give life to systems that are not easily controllable.

Indeed, if quick connections, on one hand, are making work relationships, social relationships and the like easier, on the other hand, they are distancing the relationships between places and people, creating new (non-geographical) distances.

The infrastructure that the digital city has accustomed to using in different ways has generated the root of what is now called the "intelligent city" or "smart city", in which technologies in their many forms are at the service of the city in order to both to improve the quality of life of its users and to guarantee its sustainability.

Research and practice in the field of infrastructural and urban virtualization, and not only, are today strongly oriented towards the use of smart technologies (Duany et al., 2010; de Waal, 2014; Campbell, 2012; Hatzelhoffer et al., 2012). The ability to control every action, service, desire and movement from a digital device has the aim of speeding up times and improving results. The new places of contemporary metropolises are changing their real identity in favour of an identity that can be quickly captured by a smartphone or a smart tablet. More and more sophisticated apps are offered to visitors and common users of the places, creating accelerated visions of culture and services.

Furthermore, physical space has today become (Ratti, 2014, p. 80) a counterpart of vital importance for digital existence: the use of the network has led to a different way of conceiving citizens' participation in the construction process of the territory. *Community hubs*, *network thinking* and *social networks* are some of the terms that indicate the current ways of relating to the network by communities.

Advanced technology infrastructure introduces a new culture of exchanges by creating new opportunities for interaction within the settled communities. The tension that can arise presents outcomes that may not be foreseen and have unexpected characteristics of territorialization and contextualization. Through the innovative contribution of this infrastructure, the forms of land use are being re-organized, creating contexts where social exchange replaces those that had formed over time. The processes are not easily identifiable because the signs that form do not always help with understanding the transformations that are taking place.

Such virtual communities have the common goal of supporting specialized social groups often separated by geographical distance (Graham and Marvin, 2001). In fact, new information technologies allow the generation of networks in a very short time through which urban sociality can exchange information, participate in choices and express consent.

The interaction between virtual communities and urban places has different effects. Graham and Marvin observe two elements in particular in this regard: the relationship modalities between virtual communities and the public sphere and the implications of the birth of entirely virtual cities. As Patsy Healey (1995, p. 277) asserted,

> the physical city must be replaced with a virtual urbanity, a city of the mind created by telematics'. The hope of this new technological revolution is that it will provide the channels through which knowledge and information will be more democratic, dispersed due to the diversity of relational networks in urban regions.

But there is also the risk that virtual communities could lead to further urban fragmentation and a removal of sociality from urban places in favour of electronic networks. And "ethnic groups will gather in their virtual communities, that libertarians speak only to other libertarians ... inevitably, the effect will be to demolish local geographic communities and ultimately weaken the entire national community" (Brown, 1994, p. 33).

Even in this eventuality,

> the widespread adoption of such sophisticated electronic spaces could lead to a further intensification of the separation of the places of the space and blur the boundary between real and simulated, between public and private. ... It is difficult to speculate on the ways in which such changes will affect cities, but there seems to be no doubt that the elaboration of a whole complex of virtual realities within which considerable portions of the urban population will spend considerable chunks of time, it will alter the ways in which they approach and use the physical spaces of cities and the nature of social interactions.
>
> (Graham and Marvin, 2001, pp. 291–292)

Technologies are, in fact, restoring dignity to the "place" as a result of the interaction between space and sociality. For this purpose, elements of tangible and intangible heritage are put in place that can contribute to increasing the attractiveness of the city, creating an experiential vision of paths and parts of it. This vision is favoured by the presence of a quality urban environment and a creative class (Florida, 2002). The smart experiential paths are an example of the use of information and communication technology (ICT) as a tool capable of supporting knowledge of the place through multimedia and interactive devices.

The use of public spaces should also be considered in this direction. Access to the network of squares, streets and parks has been made possible thanks to wireless and advanced smartphone networks and is having multiple positive effects. Recent studies relating to the influence of the internet on the use of open spaces (Hampton et al., 2010) show how the network has expanded their use by revitalizing them, repopulating them, improving their security and contributing to greater social cohesion. The presence of totems, info points and multimedia panels constitutes a further element that extends the use of public spaces, creating the possibility not only of knowing the history of the places but also of the available services and equipment or giving the opportunity to respond with a message to questions or comments placed on multimedia panels, helping increase the interaction between users and places. Furthermore, being able to have access to multiple types of information, including those on the healthiness of places, news, social and political information, wireless favours possible exchanges of ideas among the users of the places, further affirming the public function of the space.

The analysis of the different types of networks and their connections, interconnections and overlaps are directed towards reading and interpreting, as well as planning, a metropolitan territory beyond administrative borders and beyond the traditional way of designing cities and landscapes (Sepe, 2015a).

The management of the enormous amount of data coming from the network requires in this regard a revision of the planning models and the tools with which these are implemented, as well as the definition of new boundaries in order to create adequate practices and protocols to apply them. In this sense, new possibilities arise for territorial projects capable of being in line with the contemporary scenarios of flows and networks of places (Castells, 1989; Hall and Preston, 1988; Kaika and Swyngedouw, 2010; Keith et al., 2010).

The encounter between the different spaces of flows will result in a variety of environments, generating integrated platforms that will redesign connections, hierarchies and development opportunities.

Indeed, a spatial configuration capable of favouring a complex interconnection should perform multiple functions, namely facilitating the exchange between the different modes of mobility, even soft; placing new functions in infrastructural nodes; and ensure the high quality of pedestrian, cycle, river and green paths to facilitate the presence of more typically urban functions. Furthermore, the more strictly functional aspect should be adequately connected to the high symbolic and cultural quality of the spaces attached to the infrastructural nodes, in order to create a sense of belonging and to safeguard the identity of the places.

Blue, green, cycle, pedestrian, experiential, cultural, virtual and multimedia paths can overlap, intertwine and recombine, having the potential to create new mixed urbanities that are more liveable and sustainable.

In this perspective, the project of a place that wants to act as a generator of its cultural armour must be able to propose territorial devices that act on its metabolism by stimulating its connective social, economic, educational, health and landscape functions. In this new dimension, one of the most exciting challenges is a design and planning based on the multiple creative dimensions of development able to seize the opportunities of the alliance between heritage and creativity, heritage and innovation (Carta, 2014) and heritage and health.

# 5 Indicators

The indicators created to measure aspects relating to sustainability in its various forms and meanings – such as happiness, health, liveability and resilience – are manifold (Badland et al., 2014; Boyko et al., 2012; Lawrence, 2008; Lowe et al., 2015; Pineo et al., 2018; Wong, 2006). Here four of these indicators will be reported, namely the Happy City Index, the Urban Health Index, the Liveability Index and the City Resilience Index.

Of these indicators, which have been chosen as emblematic for the purposes of this volume, the types of data considered priorities and collected for their calculation are reported in particular.

The Happy City Index (Lewis Sam, 2016) was developed by Happy City and the New Economics Foundation (NEF) to monitor the conditions that create sustainable well-being, giving people, communities and cities equal opportunities to improve with respect to present and future generations.

The Happy City Index was the first index to measure the well-being of a city in the UK. It concerns the measurement of urban well-being through the combination of external and internal factors. External factors are the drivers of well-being, namely work, education, health (physical and mental), the place (home, public spaces, transport) and culture (participation, cultural liveliness, social isolation). Internal factors include satisfaction, goals, vitality, a sense of belonging, esteem and social relationships.

The index relates economic and non-economic data, in order to connect data concerning the economic potential of a city and social benefits. Greater well-being is associated, for example, with sociability, better health and greater productivity, while good urban liveability is a factor that is increasingly taken into consideration by companies when deciding where to invest.

The structure of the Happy City Index is based on three dimensions: city conditions, equality – assessed using indicators on inequalities relating to income, health and well-being of the city – and sustainability – assessed using indicators on $CO_2$ emissions and energy consumption levels. The City Conditions dimension is deepened by considering five areas of well-being, namely work, health, education, place and community. Additional indicators – collected from national data sources and continuously updated – are added to these to frame the state of well-being of the cities and take actions for its improvement.

DOI: 10.4324/9781003098775-7

This index contains 10 design principles, and in particular, the index measures the results achieved by certain policies or loans, not the inputs offered; measures the conditions that create well-being, such as health, quality work, accommodation, neighbourhood safety, and not directly the personal well-being; integrates equity and sustainability of the well-being factors provided by cities; considers a wide range of indicators for each policy issue; recognizes a multidimensional value of the concept of well-being; considers the gaps that exist in the calculation of different indicators; integrates objective and subjective indicators; updates data constantly to record mutations in real time; collects data from local authorities and lower levels; collects data for major cities in England, including London; and unifies the data relating to urban and rural areas.

The pilot case of Bristol, which evaluated the actions carried out for the period in which Bristol was a European Green Capital, for example, showed the various types of well-being, which constitute useful data in the context of the city's decision-making processes. Among the results of the analysis of the data collected, it was found that there are some factors that have a greater impact on people's well-being than others and the identification of these factors can help establish which are the priorities with respect to services, neighbourhood or street equipment.

Accordingly, the policies that aim at well-being must be implemented taking into consideration the various alternatives that have an impact on the well-being of people.

The Urban Health Index (World Health Organization, 2014) is a metric for measuring and mapping health disparities in urban areas. The index has a flexible approach to the presentation of health data that can be provided with various types of information, including visual information, graphics and statistics on health indicators.

This index is devoted to researchers, academics, public health operators, evaluators and all figures interested in evaluating critical situations and planning improvements in the health sectors of a given urban area.

The construction of the Urban Health Index involves the combination of a set of variables for health indicators that are standardized or normalized to be comparable and combined together.

Once the values of the standardized indicators have been obtained, these are amalgamated into a single index – the Urban Health Index – calculated for each unit by taking the geometric mean of the values of the standardized indicators for each unit.

The most important aspect concerns the selection of the indicators that must be the best for the purpose and the choice is a local decision that can change over time for different circumstances and stakeholders involved. Indeed, one of the main characteristics of this index is flexibility and the possibility of modification.

The choice of addressing the index to urban areas does not preclude its application to rural areas, even if, considering its application to a small geographical area, it is more applicable to a population of dense urban areas than of sparsely inhabited rural areas, as, for these latter, a larger area of analysis is required.

The data that can be used to measure urban health are classified as follows. Regarding health, these include disease, condition, disability, health care distribution, access and cost and immunization.

Regarding the environment, useful information includes infrastructure, built environment, transportation, water quality, air quality, contaminants and noise. Regarding geography, the data may concern land use, natural barriers, population density, neighbourhood configuration, boundaries and spatial patterns. Information on the economy relates to occupation, employment patterns, unemployment, income, wages, services, economic growth and natural resources. Sociodemographic data relates to population, growth, migration, natality, mortality and education.

To calculate this index, therefore, an integrated set of data is used that frames the multifactorial nature of health disparities in order to direct decision-makers towards the implementation of global and inclusive policies. The principles for guiding the choices that are the background to the result of the indicator include the improvement of health disparities has positive effects on the whole of society, not only for those who have situations of hardship; it is important to have a long-term vision; the basis of consensus on the results and actions to be taken must be broadened; euphemism should be used as a presentation strategy.

The Liveability Index (EIU, 2021) stems from the idea that the concept of liveability evaluates the cities where there are the best or worst living conditions, assigning a score to 30 qualitative and quantitative factors with respect to five categories: stability, health care, culture and environment, education and infrastructure.

The calculation includes, as mentioned earlier, qualitative indicators, which are assessed by an economist team of experts, internal analysts and city contributors and quantitative data, which are calculated on the basis of external data.

The data collected include the availability of private health care, the quality of private health care, the availability of public health care, the quality of public health care, humidity/temperature rating, the discomfort of climate to travellers, sporting availability, cultural availability, food and drink, the quality of private education, the quality of public education indicators, the quality of the road network, the quality of public transport, the availability of good-quality housing, the quality of energy provision and the quality of telecommunications.

In order to include data relating to the period of the COVID-19 pandemic, some new indicators have been added to understand the restrictions that have been introduced in cities, namely stress on health resources; restrictions on local sporting events; restrictions on theatre; restrictions on classical and modern music concerts; restrictions on restaurants, bars, cafes and discos; and restrictions on educational establishments. The scores and effects of these indicators have been included in the health care, culture and environment and education categories.

For each category, a weight is assigned with respect to the total calculation of the Liveability Index: stability is assigned a weight of 25%; health, 20%; culture and the environment, 25%; education, 10%; and, finally, infrastructures, 20%.

The scores, then weighted to produce a uniform evaluation, range from 1 to 100, and the indicator is calculated both as an overall score and as a score for each category.

The score from 80 to 100 indicates that the quality of life is optimal; from 70 to 80, that daily life is fine but that there are some aspects that can have problems; 60 to 70, that there are negative factors that impact daily life; from 50 to 60, that there are many constraints; 50 or less, that most aspects of life are limited. The pandemic period scores for different categories have changed globally, decreasing by 7 points from the pre-pandemic score average, as the extent to which cities handled the health crisis led to changes in the ranking.

The City Resilience Index, developed by Arup and supported by the Rockefeller Foundation (2019), was created starting from the consideration that resilience in the urban environment is based, on one hand, on the acceptance of the possibility that many types of events can occur in an unsuspecting way and, on the other hand, on improving the performance of a system in the presence of multiple risks (EIU, 2021).

Accordingly, the City Resilience Index measures the performance of cities by considering four dimensions: health and well-being, economy and society, infrastructure and environment and leadership and strategy. These four key dimensions define resilience as a factor that can determine and contribute to creating health and well-being for all citizens, social and financial systems that guarantee populations a peaceful life and collective action, artificial and natural systems to protect and implement connecting citizens and inclusive and integrated decision-making processes.

Underlying these dimensions are 12 goals that cities should pursue to achieve resilience and that are key to addressing chronic issues or sudden disasters. These are minimal human vulnerability, diverse livelihood and employment, effective safeguards to human health and life, collective identity and community support, comprehensive security and rule of law, sustainable economy, reduced exposure and fragility, effective provision of critical services, reliable mobility and communications, effective leadership and management, empowered stakeholders and integrated development planning. Each goal can be achieved differently in each city.

The 12 objectives are then further specified by 52 indicators that identify the important factors for the resilience of urban systems, which include safe and affordable housing, diverse protection of livelihoods following a shock, robust public health systems, cohesive communities, strong city-wide identity and culture, flexible infrastructure services, reliable communications technology, effective mechanisms for communities to engage with government, comprehensive city monitoring and data management, consultative planning process, appropriate land use and zoning and robust planning approval process.

The prevention of health and human life implies the presence of integrated quality health services for the prevention of diseases and the protection of the population both in everyday life and during emergencies. Collective identity and mutual support imply that communities are supported by the city government and are linked in order to create a strong identity from below. This allows the communities to be able to support each other and deal with the unexpected, even in case of unforeseen events. An integrated approach to safety and law enforcement ensures a healthy population by preventing and reducing crime during stressful situations. A sustainable economic system is essential for maintaining the city's infrastructure, providing for its communities and for creating emergency funds in times of crisis and unforeseen circumstances. The quality of the infrastructure and ecosystems that protect against natural hazards is important because it can guarantee essential services, such as electricity distribution, water supply, the flow of goods, people and information continuously even in times of shock. The conservation of environmental assets is important to protect homes, offices and daily infrastructures from floods and other calamitous events. The protective function of the infrastructure must be thought about during the design phase but must also be implemented together with adequate information and understanding on the part of the population about the risks to which the city is exposed and the protection strategies that the administration intends to undertake. The quality of ecosystem services is another important factor for the functioning of the city and must therefore be achieved and maintained through active management, the diversity of supplies and the maintenance of essential public utility services together with adequate emergency planning. Mobility and communication are important aspects to ensure the circulation of information and support evacuation during emergencies, through adequate technological networks and multimodal transport systems. Effective leadership and adequate urban management must be characterized by inclusive governance, through the involvement of the government, stakeholders and civil society. Clear leadership promotes trust and is based on multi-stakeholder consultation and a process of sharing information with the population. Finally, the integrated plans that also address emergency issues such as climate change and emergency response are fundamental for dealing with crises of various types with suitable tools.

The indicators also integrate seven qualities that are identified as fundamental by Arup for resilience, namely integrated, inclusive, reflective, resourceful, robust, redundant and flexible.

The integration between systems requires that all investments and decisions go in one direction, allowing for an exchange of information between systems and faster responses due to faster feedback loops.

Inclusion requires the broad participation of communities with an inclusive approach to reach a common vision when coping with a situation of stress or shock.

Reflective systems accept the current uncertainty and modify their rules with respect to current urban conditions instead of having permanent solutions. To this end, institutions review past experiences to plan future decision-making.

Resourcefulness is a quality of systems that implies the ability of institutions to act quickly and in different ways to meet their needs during periods of shock or stress. This can be achieved by establishing priorities, investing more human and financial resources or making adequate predictions about future conditions.

The robust systems have adequate resources to withstand the impacts of various types of calamitous events without major damage. A robust design can predict disasters and give an effective response.

Redundancy is a quality that concerns a reserve created in systems so that they can both respond to extreme pressures and include the presence of multiple ways to perform a given function.

Flexibility requires systems to be able to change and adapt to new needs; this can be achieved in different ways which may require new knowledge and technologies, new practices or practices that already exist but are used in an innovative way.

The health and well-being dimension concerns the inhabitants of the city and considers the extent to which this is able to guarantee the satisfaction of basic needs in times of crisis and the health of the population with health services in normal and emergency periods.

The slightest human vulnerability allows the population to reach a basic level of well-being so as to allow people to cope even with unforeseen circumstances.

Different professions and employment opportunities allow the population to have different options and respond to change in a proactive way.

Field verification of the index was carried out in six geographically different cities – Cali, Colombia; Concepción, Chile; New Orleans, USA; Cape Town, South Africa; Surat, India; and Semarang, Indonesia – chosen because they have experienced a severe shock or period of stress.

The scores of the quantitative and qualitative indicators, generated from a normalized average of the scores, are rated from 1 to 5 on a performance scale, ranging from very poor to excellent.

Finally, the collection of quantitative and qualitative indicators has made it possible for these cities to use the City Resilience Index to understand, monitor and strengthen performance levels and modify urban actions and policies where necessary to strengthen resilience.

# Part II

# Mapping healthy and liveable places

# 6 15-Minute City

The 15-Minute City is a city able to offer all its inhabitants everything they need to live, work and have fun to be reached on foot in no more than 15 minutes.

The starting point of the 15-Minute City model, conceived by Carlos Moreno (2020), professor at IAE Paris – Panthéon Sorbonne University, is that it needs to move from urban planning to planning urban life. This can be achieved by creating a 15-Minute City in a compact area accessible to all and where people can live, work, take care, refuel, have fun and educate themselves. To preserve the quality of life for Moreno, two components must be changed, namely time and space. To this end, a new urban planning must be created with other rhythms of life and ways of using the urban space in order to access the main functions of the city; it is a question of transforming the monofunctional urban space into a polycentric connected city with public spaces accessible to all.

In this way, active mobility on foot or by bicycle can be promoted by reducing the use of the car and long journeys by public transport. All the actors of urban life are involved in this transformation as for its realization it is necessary to connect the demands of the inhabitants with the offer through different actions: creating an adequate mix of social, economic and cultural functions; increasing public spaces of meeting; and transforming the streets into spaces of free movement for walking or cycling and rediscovering hyper-proximity.

The origin of this approach can be found in the 1960s, when Jane Jacobs (1961) opposed the functionalist city and supported a city with a mixture in all its forms.

Proximity – as Jane Jacobs states – allows people to exchange knowledge, ideas, ways of thinking to generate diversity; when people come together, engaging environments are created, and humanity's greatest achievements are attained. Streets and other types of public spaces are the places where it is most possible to create proximity and it is important that as many people as possible can use them. It is necessary to create the right conditions to

DOI: 10.4324/9781003098775-9

make a place accessible to everyone as each place is different from the other; the diversity of the city is also based on the fact that so many people with different tastes, needs and abilities are close to each other.

Jacobs pays particular attention to the role of the streets in the city, stating that these are the most vital organs; lively and attractive streets make the city interesting. The conditions for diversity in the streets are that different parts of a district must perform more than one primary function and must ensure the presence of people with different hours who can use many facilities in common; that roads need to be well connected to each other in order to generate interest and further uses; that the district must provide services, activities and goods with a different range of prices, especially accessible; and that there must be a dense concentration of people who are in that place for different purposes. The combination of these conditions results in vibrant districts and cities with – at the same time – good economic effects.

In continuity with these ideas, the city of 15 minutes offers more and new possibilities of use within the same place: a cinema can become a laboratory of ideas in the morning, a nightclub can act as a gym in the afternoon and bring art and culture on the street with digital cultural platforms or to raise awareness of historical traditions with multimedia technologies. The desired city is in solidarity and participation where there is exchange, creation and mixing.

Accordingly, the way of living and staying in the city changes thanks to the "multifunctional features" that allow a reduction in travel to reach all spaces, making proximity accessible even to the most fragile (Moreno, 2020).

The principles of the 15-Minute City therefore include residents must have ease of access to goods and services; each neighbourhood must have a variety of housing types in terms of size and accessibility levels, including economic ones, and be close to the workplace of the people; the residents have to breathe clean air; and more people can work close to home or remotely thanks to the diversity of the work spaces available (smaller offices, co-working spaces, etc.).

To achieve proximity, city plans, regulations and zoning need to be updated to ensure public services, infrastructures and are public spaces accessible to all, as well as schools, small health facilities and retail, including fresh produce shops and pharmacies. This could be achieved by opening shops on the ground floor of the main streets, expanding school courtyards, transforming abandoned lots and street spaces into small gardens or pocket parks and constructing small-scale offices and light-industry spaces.

Flexible use can concern buildings and public spaces, and, in the longer term, cities can encourage the design of reversible buildings that are designed to be easily converted for different uses by decreasing demolitions and reconstructions with economic benefits and in terms of greenhouse gas emissions (C40 Cities Climate Leadership Group, C40 Knowledge Hub, 2020).

It is therefore necessary to territorialize the short-distance multi-centralities by creating urban commons where residents can live, work, have fun and educate themselves and where elements such as water, air, time and space are at the centre of new urban issues. The COVID-19 pandemic for example, which had involved the world globally, on one hand imposed the restriction of confinement to avoid the spread of the virus; on the other hand, it focused attention on other ways of being present over time.

Two other important elements in this sense are chronotopia and topophilia. The first captures the rules of the common life of places of socialization, making them multipurpose and flexible and making a city of proximity and short distances rediscoverable. In this way, it is possible to understand that other times may exist, such as that of creativity and invisible resources, that become other moments and other opportunities to be seized thanks to the proximity and accessibility that derive from them (Moreno, 2020).

The city of 15 minutes offers the possibility of transforming any place into a multiplicity of possibilities where multiple uses can be thought of in each place and each of these uses can generate new creativity. The result – according to Moreno's theory – is the love for the places where people live and frequent each other, or the topophilia that allows having affections towards places and leads to respect for them and for objects.

Crono-urbanism, cronotopia and topophilia are the three concepts that allow a new way of experiencing places through a path that is not instantaneous, but which leads to a different look at the time of life.

In the 15-Minute City, the way of experiencing the city also changes: we move from suffering mobility to chosen mobility, hyper-proximity leads to new socio-economic models and plants and biodiversity become an integral part of this approach. Density must, in fact, be understood in a reasoned way with green connectors to be made with living materials present throughout the life cycle to be combined with internal green spaces, roofs, squares and streets that foster social bonds. Green spaces in the compact city are an attractive factor as they are places where people can take a break near their homes by reducing forced mobility.

Another important factor that Moreno introduces is the issue of managing water resources, which is also a change in attitude towards the life cycle of city water. The local network becomes strategic when there is a convergence between vegetation, nature and water "projected into the urban transition" contributing to the improvement of the quality of social life.

For Moreno, the services available must then be optimized with the use of digital technologies and collaborative models, transforming digitization into an inclusion factor and creating a new generation of multimodal public services.

Technological innovation can contribute profoundly to the improvement of services for citizens: the multitude of data available can become social objects and be designed for new uses in which new technologies and social inclusion are linked: from personalized public health to open spaces for culture, from open collaborative systems to the best quality of life for the third and fourth age.

Some cities already have visions and plans for the 15-Minute City. These include Paris, Portland, Barcelona and Melbourne (C40 Cities Climate Leadership Group, C40 Knowledge Hub, 2020).

Paris was the first to adopt the 15-Minute City approach, using design to decrease the time lost to commuting and improve air quality to achieve zero emissions by 2050. For this purpose, cycle paths and pedestrian paths have been expanded, and the aim is to transform the city into a set of neighbourhoods where it is possible to find everything within 15 minutes on foot or by bicycle from home. Specifically, the plans include a cycle path on each street, the use of street parking for other uses, the creation of co-working spaces and the encouragement of people to use local shops and the creation of small parks in schools that remain open for the local population after school hours.

Portland's 2015 Climate Action Plan (City of Portland, 2015) sets the neighbourhood's goal that 80% of residents by 2030 can easily reach and access all their daily non-work needs on foot or by bicycle. In this sense, the indicators include the distance from cycle paths, the neighbourhood park and services and the good quality of the sidewalks. The Portland plan prioritizes low-income neighbourhoods to achieve their complete improvement.

Barcelona has created a system of superblocks – *supemanzanas* – which modifies the road networks of blocks of 400 × 400 meters in order to create pedestrian and bicycle connections and improve the availability and quality of spaces for leisure and community activities. The internal roads have a speed limit for vehicles of 10 km/h, the space devoted to the circulation and parking of cars is reduced and the logistic platforms for goods are moved to the edges of the blocks. This way priority is given to pedestrians and cyclists and traffic remains on the roads at the edge of the superblocks. Each choice is made in collaboration with the residents and the measures adopted are reversible in case of internal changes to the block that require it (Ajuntament de Barcelona, 2013).

The Melbourne Plan for 2017–2050 (2017) is based on the 20-minute neighbourhood, which translates into giving people the ability to reach the activities and services they need within a 20-minute walk from home or by bike or with the possibility of safe transport. The strategy in this sense is aimed at ensuring that policies encourage services and jobs close to housing. These neighbourhoods are thought of as well-designed places, with a mix of uses and homes, connected and walkable and where people can live, work and have fun.

The peculiarities of a 20-minute neighbourhood include being safe and accessible for both pedestrians and cyclists, offering quality public spaces, facilitating access to quality public transportation capable of connecting people to job services, providing housing that makes local transport and services sustainable and favouring local economies.

Finally, the rediscovery of proximity translates into resilience in the face of crises that can manifest themselves over time (environmental, pandemic, etc.), moving towards the principles of sustainable development and the protection of urban health. Polycentric life means urban and territorial resilience with cities within a short distance at all territorial scales (Moreno, 2020).

# 7 Flexible city

The flexible city is based on tools for architectural and urban planning and design, including financial and legal tools, which are not rigid and able to allow changes in the course of implementation of projects, in case of unforeseen events or new requirements and needs.

The principles of the flexible city should be related to the entire life span of an urban project, starting from the early stages. An urban project is carried out in the medium and long term, and it is necessary for its sustainability and the reduction of risks to foresee that use, times and design are flexible. Over 30 years of time, it may happen that a commercial land becomes a residential building, or a design phase may be delayed, requiring adjustments to the master plan that can only be made in the presence of a flexible tool designed for places where people can work and live well. Architects, urban planners, legal professionals and developers, as well as the various actors involved in different ways in the design and construction process of the site or building, are required to design in an innovative way, offering the greatest possible degree of flexibility (Friedman, 1997).

Flexible planning makes it possible to use local peculiarities as a starting point for the redevelopment project, resulting in lower investment costs, shorter construction times and greater support from stakeholders. The various design possibilities that can be realized with respect to the existing situation are identified through quantitative and qualitative analyses, of which the latter determine the implementation of flexible planning (Bergevoet and Van Tuijl, 2016).

Studies on the flexible city have been carried out by many architects and urban planners since the end of the 1990s and early in the 2000s for what concerns the housing models (Towers, 2005) and taken up on a larger scale by Bergevoet and Van Tuijl (2016) and CRA Associati (2018) and several other scholars. Here we report the idea of a flexible city by Bergevoet, Van Tuijl and Ratti with the case studies of the Navez-Portaels district contract in Brussels, Prinzessinnengarten in Berlin, and The Dynamic Street by Carlo Ratti.

The realization of the redevelopment projects of the places are determined (Bergevoet and van Tuijl, 2016) by four types of factors: the organizational factor which is determined by the local stakeholders, which include the traditional ones such as developers and public administrations, and all the other

DOI: 10.4324/9781003098775-10

subjects involved in the operation up to the users; the legal factor that depends on structural plans, zoning plans, local regulations and standards; the financial factor that is conditioned by the values of existing properties and by the income that can be obtained from leases, financing and refinancing; and the spatial factor that depends on the physical and technical characteristics of the place, as well as on the historical ones.

Flexibility based on use is based on concrete requirements, not on speculative ones, and on the fact that all parties make decisions together. In this way, not limiting the choices to just a small group of professionals and ensuring that demand develops only for the space where there is actual need, the investment risk is reduced.

To achieve this, it is necessary to obtain permits for flexible planning in agreement with all interested parties, even if not all of them will have the same weight in the decision-making process. Their influence will depend on the project, and in general, the owner and the initiator have greater importance in determining the decisions to be made, on the subjects to be involved and on times and methods. It can be expected that in large-scale projects, the public interest will be greater, while in small-scale ones, the parties involved will have greater influence. Architects, planners and construction experts in the costs list the different stakeholder steps and the parties involved take note of all this information. Then all the subjects take a joint decision with respect to tasks and responsibilities with respect to the plan, a decision which remains non-revocable for a limited period.

In the next steps, initial scenarios are defined and determine the choices in the short term with respect to organizational, legal, financial and spatial aspects (Bergevoet and van Tuijl, 2016).

The planning of the initial scenarios only allows working on further development during the execution of the project, in order to be ready for changes in the event of market changes or new ideas that can be implemented during the project's implementation. Investments are made in several stages, taking into account any unforeseen circumstances and distributing the risks over time. The unforeseen events can be of different types, from the increase in energy costs to the worsening of the general economic situation; if there are many unforeseen events, many additional steps will be necessary for flexible planning; if there are few unforeseen events, the design process will require fewer adjustments.

Adjustments to the initial idea are made through assessments that can be made during construction by creating new initial scenarios. Reflective sessions are used to reconsider the development process together with the owner, initiator and stakeholders and propose a recalibration. The time factor in this process of changing the initial scenario has an important weight.

Flexibility in planning must be supported by legal flexibility that is local, as well as based on use and time. Local legal flexibility depends on local regulations, which may be more open to considering changing situations than general regulations. Legal flexibility based on use concerns the possibility of making space available for events and initiatives that do not fit into generic legal frameworks. Time-based legal flexibility allows changes to be made in the event of contingencies in government policies or economies at the national level.

Financial instruments can be of great help in flexible planning; this can happen if financial models include tangible and intangible values and properties of smaller parts and are capable of dealing with unexpected events. Local financial flexibility concerns the use of tools that are not based solely on generalizations but that provide for the differences and peculiarities of the place, due to increases or losses in the value of land resulting from transformations or densification of parts of the city.

Financial flexibility relating to use is based on the adjustment of investments to actual use in order to avoid operating considering speculative values. In continuity, time-based financial flexibility concerns the opportunity to have flexible tools in order to reduce the gap between the timing of investments and those of revenue streams.

As stated by Bergevoet and van Tuijl (2016), spatial flexibility concerns the possibility of personalizing as much as possible the set of tools available to architects, urban planners and all those who are involved in the production and construction of buildings and spaces in order to adapt them to the different needs. Spatial flexibility should be local, that is adapting to the different pre-existing conditions in regeneration projects; based on use, that is achieving a fair balance between the interests of users and the interests of institutions such as those of public authorities; and related to time, that is in anticipation of extended project realization times, assigning to buildings and spaces a specific use for a short lifetime and therefore modifiable after a short period or, on the contrary, not assigning a use for a long time a function so that these locations can be used for different purposes.

A good practice in this direction is the Navez-Portaels district contract, in Brussels, Belgium, in the municipality of Schaarbeek (Bergevoet and van Tuijl, 2016).

The neighbourhood contract was created between Brussels Hoofdstedelijk Gewest, RenovaS representing the municipality of Schaarbeek and neighbourhood associations. The residents were represented by a commission of at least eight residents representative of the mix of the population present in the neighbourhood and nominated by other residents. The contract period is seven years, of which one to decide the program, four for the elaboration of plans, and one for its implementation, at the end of which a network had to be created for the autonomous maintenance of the planned action. Brussels Hoofdstedelijk Gewest and the interested municipalities contribute 5% of the project budget, while the Belgian federal government supplements this budget together with private investors.

The representatives of the administrations, together with the residents and potential investors, made an in-depth inspection of the neighbourhood to analyse the habitability of the housing, the availability of public spaces and the unemployment rate and then draw up a plan. All participated in the drafting of the plan and a PCGO – local committee for integrated development – was created to organize the debate between the actors involved, including residents. The plan has been adapted several times to the needs that emerged during the meetings, and the district is informed of the developments during the general meetings.

The idea of the initiators Brussels Hoofdstedelijk Gewest was to create, together with the residents, the recovery of public space to improve the quality of life, creating a space for slow mobility for pedestrians, cyclists, and people in wheelchairs and a large pedestrian bridge over the submerged railway line of the square, as a place for meeting and socializing. Existing buildings were also renovated, social projects were launched for various activities including that of unemployed people to help older adults and investments were made in social networks all to improve places and social cohesion, as well as attract new private investments.

Another case of time-based flexibility is that of the Prinzessinnengarten, a space of approximately 6,000 square metres located in Berlin in the Kreuzberg district. This space remained unused for many years and was used as an illegal landfill until 2009 when the residents of the Marco Clausen and Robert Shaw districts founded Nomadisches Grün to transform empty areas of the city into social gardens, creating the Prinzessinnengarten, among others. This area was cleaned up by volunteers to then plant the first crops, and today, the site hosts an urban garden with different types of organic herbs and vegetables. The gardens are all collective, there are no individual gardens and everyone can work together to create a green place: the activities include sowing, collecting seeds, processing, preserving agricultural products and developing new cultivation techniques.

The garden is privately owned by the Liegenschafts fund of Berlin and is open to all during opening hours, while it is closed at night; the income is derived from collaborations with schools and kindergartens.

The creation of the garden has increased the social cohesion of the neighbourhood which is located in a not-very-green area with social problems; the exchange of experiences between residents who learn cultivation techniques, urban ecology, sustainable consumption and new ways of experiencing places. The development of the Prinzessinnengarten has also led to a regeneration of the surrounding area with the creation of creative initiatives, co-working places, the opening of shops for sustainable design, the creation of workshops, nursery schools and cafes.

The creators of the garden also thought of a financial tool to ensure that the contributions of the participants go to pay a professional gardener, whose skills are disseminated so that they can be learned by the volunteers.

The maintenance of the site depends on residents and donors; the income from coffee in the garden and agricultural products is used to pay the rent of the land for one season. Even when the land is leased, it does not belong to the tenant but remains accessible to all. However, it may happen that an investor intends to rent the garden for a year, and in this case, it can remain closed for that period. For this reason, the founders of the garden have created a mobile vegetable garden; that is the buildings are used as a sort of container where the plants are grown in baker crates and cardboard boxes so as not to interrupt their growth (Bergevoet and van Tuijl, 2016).

Another example of a flexible approach is the Dynamic Street, a prototype created by Studio Carlo Ratti Associati with Sidewalk Labs of Alphabet, which consists of an adaptable and reconfigurable modular flooring system for the different needs of people (Cra Associati, 2018). The Dynamic Street is made up of modular hexagonal-shaped finishes that can be replaced in a short time – hours or minutes – to change the function of a street. The hexagonal modules can be integrated with lights and accommodate vertical structures including poles, bollards and baskets for basketball. This flexibility can allow a change of use in the different morning and evening hours – when there is greater use, for example, of workers who walk or cycle in the early hours and then of families in the following hours – and on holidays, at weekends, and on the occasion of events. It is also possible to imagine a changing use to create the physical distancing required by pandemics or in the first stages of post-natural disasters.

In continuity with the concept of flexibility is also the use of roads with self-driving vehicles that in the future will free up space on motorways to be used and converted for other activities, from urban agriculture to space for games, from sport to leisure (CRA Associati, 2019). Five visions of this were presented as part of "The New Deal" of the exhibition "Les Routes du Futur du Grand Paris", organized by the Forum Métropolitain du Grand Paris and inaugurated at the Pavilion de l'Arsenal in 2019 (Figures 7.1–7.3).

*Figure 7.1* Image of "The New Deal – Habiter les Lateralitè" in the framework of the exhibition "Les Routes du Futur du Grand Paris", organized by the Forum Métropolitain du Grand Paris

(Source: Photo credits by CRA-Carlo Ratti Associati)

*Figure 7.2* Image of "The New Deal – La Voie Dynamique" in the framework of the exhibition "Les Routes du Futur du Grand Paris", organized by the Forum Métropolitain du Grand Paris

(Source: Photo credits by CRA-Carlo Ratti Associati)

*Figure 7.3* Image of "The New Deal – Multimodalite Partout" in the framework of the exhibition "Les Routes du Futur du Grand Paris", organized by the Forum Métropolitain du Grand Paris

(Source: Photo credits by CRA-Carlo Ratti Associati)

Flexibility constitutes an objective, as well as a methodology, aiming to achieve sustainability – meant in its triple meaning – and reduce risks. Another aspect to take into account, as in the other examples shown in the volume, is the attention to the identity of the places, namely considering the different variables that can occur during the process of planning and designing a place; some constants, such as the identity of the place, must be taken into account in order to both protect and enhance it.

# 8 Soft city

The soft city, created by David Sim (2019), is based on the idea that from the union of density and diversity a more liveable and healthier city can be obtained, as the proximity of an environment can be translated into time. Indeed, services, public spaces and nearby transport allow to use the hours they would use to reach places not nearby in time to be able to devote to other activities.

Many activities, in fact, require the use of a car to be reached, but not everyone – children and older adults for example – can drive and often the activities within a family are different. The greatest cost concerns the time it takes to go from one place to another, and that could be dedicated to oneself and to building relationships with friends and family.

According to Sim, another important aspect of proximity is the sharing of places and resources between people, which creates a feeling of belonging in the community. The local identity can be stronger than the national or cultural one, and being more inclusive, it can be considered one of the healthiest forms of collective identity. The characteristics of comfort and connection must be combined with those of variety and proximity in order to obtain a better quality of life; every detail of the urban composition of the neighbourhood is important and must be taken care of so that every activity can be carried out in a healthy and attractive environment, capable of giving well-being and creating an opportunity for socialization. In this sense, the neighbourhood is not just a place but also a state of mind.

The way to increase the time available is to place more activities that are part of a person's day at close range so that they can work, live, learn and take a break in the same place. In this way, the hours spent by a person on travel could be reduced and be translated into savings of energy, money and, above all, time. This should be combined with the creation of more pleasant and connected places for all activities.

Sim's idea is that soft urban relationships are more sensitive and reactive and can last longer than hard ones; in this sense, the "soft is more difficult to break".

Life is constantly changing, and the urban structure must adapt together with people's lives, building organic and soft places. The soft city is a combination of hardware, which is the streets and buildings – and everything that

DOI: 10.4324/9781003098775-11

is built – and software that affects the invisible part such as behaviour, trust, planning and legislation.

The soft city can be seen in larger and smaller phenomena, explicit or subtle and which are able to make life more pleasant, because, as Gehl (2010, 2016, 2020) and Jacobs (1961) write, the urban form can influence the behaviour of people.

To measure the success of densification operations, quantitative indicators are generally used to not fully cover urban complexity; for this purpose, qualitative criteria are needed to make clear; for example how much of an urban form is able to facilitate the connection between people for greater social opportunities, or how much it facilitates access to services and places?

For this purpose, Sim identifies nine criteria, illustrated in the following, to which a liveable, resilient and high-density area should respond, namely having a diversity of built forms and outdoor spaces, having flexibility, being on a human scale, having good walkability, having a sense of control and identity, having a pleasant microclimate, having a lower carbon footprint and having greater biodiversity.

To achieve diversity of the built form, it is necessary to have different activities within a neighbourhood and many different types of buildings. The urban form must, in turn, be able to welcome this diversity of volumes and forms of buildings, and the whole should also be well balanced from a visual point of view and not prevail over nearby buildings and businesses.

The outdoor spaces should be made with an adequate variety in order to allow activities of different types, giving greater value not only to activities, such as going to the park, but also to daily activities, such as waiting for the bus or throwing out the garbage, that should be carried out in places pleasant that could be opportunities for socialization. Outdoor spaces should constitute a system of interconnected public and private places that allow for different uses.

Streets, squares and public parks offer different opportunities than private gardens and courtyards; if public spaces and private spaces can coexist close together and complement each other, greater possibilities of choice of activities could be offered to more types of people. There should also be spaces of different sizes – small and large and with varying degrees of privacy; between public and private, there are sub-categories such as semi-public and semi-private, which further expand the typology of spaces. Other types of spaces could be flexible and multipurpose spaces – where different activities can be carried out at different times of the day – spaces dedicated to sports, games and other specific activities and inside–outside spaces. The latter include porches, balconies, verandas, decks and terraces.

The streets are also of different types – main, secondary, avenues, alleys – and can create different ways of being lived and of movement, just as parks and squares could be part of a path for going to work and to a garden in a courtyard, constituting a shortcut. Standing and sitting are therefore actions as important as moving and can be thought in a wide variety of spaces.

A neighbourhood should respond to changes due to densification, new activities, new functions, new people or new needs in the short, medium and long term, and these changes could depend on the hour, day, season or weather. A school courtyard that can become a public park on holidays or a hotel lobby that can become a pop-up store are just some of the examples in this direction.

Urban spaces should be made on a human scale, that is, thinking about the comfort of the people who will have to use them and spend time there; this means designing smaller spaces and buildings with lower heights while paying attention to what can be seen at eye level. Spaces on a human scale also mean making sure that people can perceive sounds, scents and natural materials in better microclimates as well as making meeting and socializing between people more pleasant. Last but not least, the perception of safety can be better provided in smaller places, where it is possible to have an overview of the space.

The design of a place should provide for good pedestrian traffic or create a connection between people and their neighbourhood, making it convenient and pleasant to access the activities, spaces and services which are present, as well as the transition from one situation to another in a simple way. This can be achieved by creating different and protected options for moving between buildings, blocks and neighbourhoods, that is by creating suitable positions for windows, doors, corridors and stairs. The connections between the spaces also create opportunities to get to know and meet people and experience new places (Figure 8.1).

*Figure 8.1* Vision of the soft city

(Source: Jan Gehl, Archive's)

Urban spaces should consist of clearly recognizable and physically defined places, owned or controlled by a person or a group of people. The spaces can be very small, such as a step in front of a door, where people can put a plant; or a courtyard with a staircase in common with eight families, where people know each other and can exchange advice or help in the event of a small emergency; or a public square accessible to all. The important thing is that these internal and external spaces are identifiable and recognizable on a small scale and on a larger scale.

The physical comfort that is created with a good microclimate affects people's lives because it offers greater opportunities to spend time outdoors and perform various activities such as walking, cycling, going to the park and using public transport. Improving the microclimate, in continuity with the creation of differentiated spaces, contributes to improving people's quality of life. The reduced height of buildings and small spaces can have an impact on the environment reduced to a minimum, as they improve local microclimates. Less exposure to wind and sun also reduces the need for artificial heating and cooling and maintenance costs.

Biodiversity must be understood in a holistic way; therefore, including the life of vegetation and the life of animals offer beneficial effects on built environments.

A rich and diverse nature reduces people's stress as it can help reduce the noise of means of transport by absorbing them, can reduce smog by purifying the air, can create a filter to protect from the wind and the sun and can create greater privacy.

Furthermore, as Souter-Brown (2014) states, the natural environment and landscape can be perceived differently and provoke positive sensations in people, especially if they are conceived of as landscapes on a human scale: they are "slower" places where it is possible to perceive the scent of flowers, provide natural shade and restore biodiversity; in this sense, the small-scale garden and the larger-scale nature trail are examples.

Finally, the concept of a soft city can also be extended to the use of technologies and where different types of information can be offered to citizens in real time (Skelton, 2014), as mentioned in Chapter 9 on the smart city.

The design of soft places in this sense never ends because as communities change, so do buildings and their public spaces, physical and digital, as well as uses and habits.

# 9 Smart city

The infrastructure we have become accustomed to using in a different manner in the cyber city is at the heart of the smart city in which the whole range of technologies are at the service of the place both to improve its liveability and healthiness and ensure its sustainability. In this case, the hybridizations vary widely in terms of both significance and application so that the issues are at times hard to define (Duany et al., 2010; Karvonen et al., 2019; Mitchell, 1995).

Indeed, in recent years, the social sciences have been adopting the perspective of the relationship between health conditions and economic conditions in certain geographical-environmental contexts. The quality of life based on place has an important impact on people's health and happiness, just as the appreciation of the latter in relation to a neighbourhood or city depends on the built environment and social life. In order to be able to clearly assess the effects of places on people, a huge amount of data is needed to be linked with emerging digital technologies.

Furthermore, quality of life encompasses a wide variety of both objective and subjective characteristics that are part of the urban system. Factors that have been particularly highlighted in recent decades include health, green spaces, cultural services, happiness and urban well-being (Sassen and Kourtit, 2021).

As Sassen and Kourtit (2021) state, in order to achieve an urban transformation based on sustainability in its broadest meaning and to provide an adequate impact on the quality of life, it is necessary to understand which city governance is important to face the new urban threats. This also refers to a smart transformation with integrated data management to be used in short- or long-term crises, as happened for the COVID-19 pandemic.

This pandemic has brought about a notable change in people's movements, and in the way, they view urban life, and this changed situation has led to the question of how intelligent transformation can be achieved with sustainable construction of cities in times of health emergencies.

DOI: 10.4324/9781003098775-12

The potential offered by digital technologies is manifold and includes the analysis of urban data (e.g. those coming from social media and big data); management of public space, with the use of multi-actor multi-criteria analysis MCA models or urban 'body and soul' analysis, projects of safe and environmentally friendly environments; devices for the online security of public spaces for the safety of places and the management of environmental disasters; forms of e-democracy for new ways of participation among citizens; systems for early warning relating to maintenance activities; and use of social technologies for city imagineering.

In times of crisis, smart cities can give added value also from a socio-economic point of view by creating policies that include the intelligent management of public space with intelligent mobility plans, urban traffic lights and smart street lighting, urban furnishings and flooring made with new materials.

It can therefore be said that intelligent cities are organisms made of "complex, dynamic and interacting spatial bodies". As Neal (2012) states, "they are living intelligent organisms with complex, dynamic and interacting spatial 'bodies' providing a mix of multiple opportunities, dreams, social interaction options and alternatives and strategic and personal choices to make, in a dynamic, sometimes global network society".

This new way of thinking aims to improve urban structures and infrastructures to make cities smarter, more liveable, healthier and happier with a high quality of life for all. Putting people first in the reference protocols of smart cities translates into considering end users with their needs as a priority, involving them in forms of collaborative partnerships. In fact, digital technologies are not objectives but tools for obtaining data to improve the quality of spaces through the representation of a balanced mix of information offered from a multifocal perspective (Sassen and Kourtit, 2021).

The concept of a smart place is strictly connected to that of soft place as shown in the previous chapter. In this sense, communities are the soft part of the city, which, thanks to online media, can be in contact even if they change neighbourhoods.

A further and increasingly current possibility is represented by a mixed virtual/tangible reality that leads to new forms of citizenship that meet different needs (Skelton, 2014). Similarly, it is already happening on the smallest scale in a single room, in a building and on a campus, where hybrid forms of reality already exist and where physical structures and a considerable amount of electrical and communication cables and virtual spaces coexist with Wi-Fi connections. And the configuration of many mixed places is constantly varying as the uses of buildings and spaces are constantly changing.

Currently, in Europe, smart cities are concerned with projects for retrofitting historic cities, in which sensors are placed in the urban environment to collect data that can be used by citizens, including energy consumption, the level of carbon dioxide and the use of services (Potesta, 2019).

Smart cities are proposed as cities that have objectives of increasing liveability; many companies that market digital platforms, for example, advertise their products stating that these are able to improve the lives of citizens as a better connection corresponds to a better quality of life. However, it is not easy to calculate in what percentage a good connection can contribute to a better quality of life and how much other factors contribute.

The indicators for the quality of life are calculated considering the factors that determine it. The Mercer Quality of Living Survey, for example, evaluates services by ranking them with low, medium and high scores, corresponding to the degree of happiness they are able to provide to citizens. The efficiency of services in this indicator is considered a primary factor.

The smart city is based on the idea that its services are accessible to all; urban living does not therefore involve sharing a place but sharing a network of services. The communications made through the internet allow everyone to have a better quality of life as the connections are not based on the physicality of the place and can therefore also take place with people living in dispersed areas. The dematerialization of space, however, leads to the potential disintegration of the place: while the interaction between people and physical space leads in most cases to a sense of place, in the case of virtual spaces, since there is no physical existence, these can rarely lead to the creation of a sense of place in the traditional meaning of the term (Potesta, 2019).

Indeed, the new urban environments resulting from regeneration processes must satisfy not only today's population but also future generations. These include the construction of infrastructures by integrating new technologies and peculiarities of the area, restoring well-designed, liveable and healthy places and spaces. To achieve intelligent placemaking, it is necessary to process and interpret all available data, an essential factor for the urban designer's toolbox (AAVV, 2018a).

The use of technologies such as social media can help urban planners to evaluate trends in real time to understand how people use cities, spaces and infrastructures with the aim of creating a place with a distinctive, healthy and liveable character.

One of the challenges is knowing how to create connected and intelligent environments using the latest technologies but, at the same time safe, easier to use and attractive, promoting both physical and mental well-being. Pedestrian places are places that offer many possibilities for interaction between people, and these interactions can be supported by technologies by connecting design and new digital tools to improve their liveability. With data analysis, it is possible to both create key attributes of the place and identify the many opportunities that the latter can offer.

Smart roads, for example, offer an integrated and multimodal network capable of improving accessibility and air quality. The physical urban development integrated with the virtual one offers many other opportunities in favour of healthy and liveable places.

In a holistic smart-city scenario, this enhances its collective intelligence as a place of creativity, inspiration, knowledge-based urban development. Likewise, technologies allow citizens to navigate digital cityscapes with an active role as both player and creator. The multiplier effect of the dynamics of social networks makes citizens narrators of the reality of places, through the sharing of images, comments and insights published on social networks.

At the same time, new technologies are leading to smart mobs, spontaneous social mobilization dynamics, digital coherograms, information and communication technology (ICT)–mediated representation of places and bottom-up initiatives connected to citizen-driven innovation. Furthermore, there are also pop-up smart cities (Townsend, 2013) that combine creativity, scientific infrastructures, well-being and entrepreneurship.

The presence of innovative technologies in a place is also becoming an element to strengthen its positive urban perception; being considered smart attracts not only more visitors as it makes the place more attractive and liveable but also more workers as they can work while continuing to stay connected with everyone in an innovative and dynamic environment.

The role of ICT in the branding of the place is increasingly wider and aims to offer an experiential character through co-creation and co-branding operations, capable of highlighting the potential and information that emerged from the stakeholders in the various digital environments in which these interact.

Tangible and intangible cultural resources that can contribute to the attractiveness of the city are linked by creating an experiential vision of dreams and paths. This vision is favoured by the presence of a quality urban environment and a creative class (Sepe, 2015a, 2015b).

The experiential paths are an example of the intelligent use of ICT as a tool to support knowledge of the territory through interactive and multimedia devices. The routes can include the use of multimedia kiosks with Wi-Fi access for tourist and cultural information via mobile apps. In this way, a system of use for cultural heritage is created, organized with a view toward innovative enhancement of the values and symbols of the territory and in a sustainable way.

Finally, another use of new technologies and the network is augmented reality: using physical and virtual tools, it is designed to enrich "human sensorial perception by means of information, generally manipulated and conveyed electronically, which would not be perceptible using the five senses" (Carta, 2017). There are, in fact, many combinations of physical and virtual that can be implemented to fulfil a range of different uses.

However, despite the various digital opportunities for improving liveability, there is a gap between the idea of smart city and its real effect on the healthy and happy city.

There is also a cause for concern related to the smart city. The presence of monitoring devices can also be a control tool, as data can be shared by citizens but also in platforms managed by individuals, determining a sort of limitation to people's freedom.

In fact, the potential of smart places is vast, and in order to be used in a sustainable way and of real help for urban health and liveability, it is necessary that new technologies are more closely connected with the peculiarities of the places and the desires and needs of citizens so that real and virtual spaces could collaborate in synergy and technologies would represent an added value to the transformation of cities and concrete support in times of crisis.

# Part III

# The Healthy Pl@ce Design method

# 10 Analysis

The original Healthy Pl@ce Design method aims to detect factors which make places healthy, liveable and happy from the user's point of view and identify appropriate project interventions to enhance or design these factors (Table 10.1; Appleyard, 1981; Burton, 2015; Burns, 2005; Ballas, 2013; Kyttä et al., 2015; Pykya et al., 2018; Sepe, 2018, 2015b, 2017b, 2019a, 2020a, 2020b; Taylor et al., 1998; UN Habitat, 2013).

The analysis part of the method – five phases – consists of different kinds of surveys, observations and questionnaires. Part of design is composed by three phases and includes checking the consistency with the 25 principles of the "Charter of Urban Health and Liveability Design", a questionnaire and the identification of the project interventions. The final results are visualized through two maps, one of analysis and one of design, which contain the synthesis and overlapping of surveys, questionnaires and collected observations.

In the following, the different phases of both analysis and design are reported in detail.

## Phase 1: Definition of the study area

Phase 1 of the method consists of defining the study area; the researcher needs to go to the site in question on different days and at different hours and decide, through one or more inspections, whether to confirm the delimitation decided beforehand or modify it.

The choice of the area to analyse constitutes the starting point of the method and is one of the most important elements. The criteria for deciding the place to survey include continuity of activities that are carried out, main urban design characteristics – including dedicated pedestrian and cycle paths, gardens and urban green, waterfront, illumination and urban furniture – and different uses.

DOI: 10.4324/9781003098775-14

*Table 10.1* Healthy Pl@ce Design method, scheme

| *Phase* | *Objective* | *Actions* | *Product* |
|---|---|---|---|
| 1 | Definition of the area | Visits on site | Definition of the borders to analyse |
| 2 | Observation of characteristics of the place | Surveys of<br>People<br>Activities<br>Perceptions (singular and mixed)<br>Elements | Definition of the characteristics of the place |
| 3 | Questionnaire to place users | Identification of elements which make that place healthy, liveable and happy from the peoples' points of view | Mosaic of user ideas |
| 4 | Traditional analysis | Identification of elements that compose the place | Mosaic of the traditional elements which compose the place |
| 5 | Identification of factors of urban health and liveability | Overlapping of all data collected | Map of urban health and liveability analysis |
| 6 | Check of the degree of urban health and liveability | Overlapping between the map of urban healthy, liveability and the 25 principles of the Charter of Urban Health and Liveability Design (UHL) | Definition of factors to be improved |
| 7 | Check of the design ideas with the users of the place | On-site questionnaire and research on the websites with the user requests | Mosaic of degree of pleasure concerning the design ideas |
| 8 | Identification of project interventions for the realization of the Charter principles | Overlapping of all data collected | Map of UHL design |

## Phase 2: People and activities

Phase 2 is characterized by observing the characteristics of the place through three surveys concerning the kinds of activities, the perceptions and the elements which contribute to the feeling of healthy, liveability and happiness of the place (Appleton, 1975; Dines et al., 2006; Kaplan and Kaplan, 1989; Maas et al., 2006; Thompson and Aspinall, 2011; Zlot and Schmid, 2005)

In survey 1, as shown in Table 10.2, the types of people – locals, visitors, professionals – and activities – enjoyment, passing by, sport, work – are observed. These activities need to be observed from the quantitative point of view to collect data concerning what percentages of the activities are present

*Table 10.2* Healthy Pl@ce Design method, survey 1: people and activities

| *Activity* | *People* | *Quantity of People* | *Frequency* | *Pace* |
|---|---|---|---|---|
| e.g. enjoyment, passing by, sport, work | e.g. locals, visitors, professionals | low, medium, or high | low, medium, or high | rapid, moderate or slow |

*Table 10.3* Healthy Pl@ce Design method, survey 2: perceptions

| | *Visual Perception* | *Sound Perception* | *Tactile Perception* | *Smell Perception* | *Taste Perception* | *Mixed Perception* |
|---|---|---|---|---|---|---|
| | e.g. the river | e.g. water of the river | e.g. grass | e.g. flowers | | e.g. comfortability |
| Quantity: low, medium, high | high | medium | high | medium | | high |
| Quality: Pleasant, annoying | pleasant | pleasant | pleasant | pleasant | | pleasant |

in that place and how they influence its liveability and healthiness. Then, these activities need to be both measured and observed regarding the presence of persons from the quantitative point of view, expressed in the low, medium, or high presence of people.

Accordingly, the frequency with which the activities are repeated or implemented and at what pace are measured: it is observed whether that activity is carried out with a rapid, moderate or slow pace.

Survey 2 consists identifying singular and mixed perceptions (Table 10.3). The singular perceptions include the visual, sound, tactile, smell and taste perceptions, while the mixed ones include those of chaos, serenity, disorder, joy, harmony, disorientation, uncomfortability and so on and depend on the sum of one or more perceptions. Their quantity is expressed as light, medium and high percentages, and the quality is expressed as a pleasant or a disturbing perceived perception.

Usually, people describe a place only through visual perception; in the reality, the perception of a site involves all five senses. At the same time, the perceptions of serenity or chaos are given by one or more sense contemporaneously, for example by a pleasant or disturbing visual, smell and sound perception. The accuracy of the perception survey, although characterized by a certain degree of subjectivity, is capable – together with the other

*Table 10.4* Healthy Pl@ce Design method, survey 3: furniture and equipment

| | *Accesses* | *Paths* | *Urban Furniture* | *Equipment* | *Means of Transport* | *Art Product* |
|---|---|---|---|---|---|---|
| Typology | e.g. open area | e.g. cycle path | e.g. light street lamps | e.g. bike racks | | e.g. sculptures |
| Peculiarity | e.g. no borders | e.g. separated by the pedestrian path | e.g. good quality design lamps | e.g. good quality design racks | | e.g. enaging sculptures |

surveys – to support identifying many factors related to the liveability, healthiness and happiness of a place.

Survey 3 of this phase consists observing the elements which contribute to the health and happiness sensation, such as constructed and natural elements, suitable pedestrian paths, sustainable means of transport, art product, good-quality equipment and urban furniture (pavement, benches, lighting, wireless, etc.) and the kinds of accesses (Table 10.4).

Many factors can contribute to the sensation of liveability: the right identification of these can help to suitably observe paths, behaviours, movements and activities of people.

Finally, from the intersection of these data, a first outcome of the degree of healthiness, happiness and liveability is obtained, resulting from the surveys on the place being analysed (Buttimer and Seamon, 1980; Carmona, 2019b; Kent, 2008; Madanipour et al., 2014; Jones, et al., 2008; Kent, 2008).

## Phase 3: The place from peoples' points of view

Phase 3 consists in a questionnaire administered to the people who use the sites aimed at identifying factors and elements which give them the sensation of a healthy, liveable and happy place.

This questionnaire is mainly administered to common users of the place being analysed, including passers-by, locals, residents and visitors of different ages and nationalities. In the case of the specific objectives of the case study, privileged witnesses and stakeholders can be interviewed to allow for further interpretation of the site.

To introduce the questions, some personal details are asked of interviewees – such as their age and if they are local or, if not, their nationality. This is useful to comprehend the perception of the place, which could change if people already know it, usually walk there or are tourists who are visiting the site in object for the first time.

Questions may include the following and will be modified in accordance with the place characteristics:

1. Does this place give you a feeling of happiness or sadness/liveability or discomfort/healthiness or unhealthiness?
2. What are the elements that give you the above sensations?
3. What are the main facilities that give quality to this place?
4. What are the activities that you do in this place, and how often?
5. What do you think about the presence of many or few people here? Do you think that it is capable to improve the pleasantness or unpleasantness of the place?
6. According to the place's healthiness, liveability and happiness, what could be done in order to improve this place?
7. What is a healthy/happy/liveable place that you remember in this city or elsewhere?
8. How might the weather conditions influence the perception of the healthiness, liveability or happiness of this place?

The observations of the place carried out during the previous surveys allow the choice of the most suitable part of the area to administer the questions. Usually, the presence of fountains, benches, steps and other kinds of urban furniture around which people stop constitute sites where people take a break and are more available to answer.

## Phase 4: The traditional analysis

Phase 4 is that of analysing the traditional cartography in order to understand the elements that compose the place in terms of the type of the historical and architectural elements, urban fabric, the natural environment (green areas, sea, hills, etc.) and other public spaces in the surrounding area. A collection of projects of urban design in development of these areas can complete this analysis.

It is important to suitably identify if the area in question – or its surroundings – is designed in the framework of a wider urban design and planning project or is a consolidated "place". This is because a new project could change the current urban liveability and health, and eventual new project interventions identified by the method could already be forecasted by the new plans.

## Phase 5: Mapping health and liveability

Phase 5 involves the construction of the map of urban health and liveability with the identification of spaces and features that give to the people who use that place the perception of these factors.

The map will be the result of the different survey operations, analysis, questionnaires and observations, which were collected on the sites in question.

To this aim, suitable symbols are created, that include single perceptions, mixed perceptions, equipment and urban furniture, pedestrian and cycle paths, historical buildings and monuments, playgrounds, sport areas and equipment, people, nature activities, green paths, green, sea–river–lake, area of environmental regeneration, architectonic and artistic elements, kiosks–bars–restaurants, info points, flying kites and urban beaches. These constitute the basic elements which are surveyed; considering the uniqueness of any place, other elements and related symbols can be added.

For these reasons, drawing the symbols is constituted by lines, basic geometric shapes and colours to allow other people who could use this method to add others.

# 11 Design

## Phase 6: The consistence with the Charter of principles

In phase 6, the check of the degree of healthy, liveability and happiness is carried out. This is obtained through the study of both intrinsic and extrinsic factors contained in the map and which are capable to determine urban health, liveability and happiness. The intrinsic factors include the perceptions, tradition and culture. Extrinsic factors include the architecture, facility, and urban furniture. The aim is that of identifying factors that can be improved according to the following 25 principles of the Urban Health and Liveability Design concerning the following Charter (see the Conclusion). The principles result from all the case studies carried out in the experiments using the method.

1. A healthy and liveable place is a space which can transmit feelings of health and liveability to everyone who uses it.

   Accordingly, it is important to do the following:
2. To encourage the use of the place by people of different age groups, from children to older adults
3. To eliminate architectural barriers which might discourage people from frequenting that space
4. To create a suitable balance between the elements of nature, landscape and equipment in the composition elements of the space
5. To have both in streets and public spaces natural lighting during the day and artificial lighting at other times, avoiding artificial light in daily hours
6. To retain an adequate state of cleanliness and maintenance
7. To create suitable spaces for dogs and domestic animals
8. To create a sense of security and safety to those who walk, cross, rest and so on in the public spaces
9. To minimize or eliminate the noise generated by public transport
10. To improve suitable cycle lines
11. To fully perceive naturally occurring smells – for example, wood, grass, sea

DOI: 10.4324/9781003098775-15

12. To have direct contact with natural materials, preferably local, used in the design of the space
13. To have the presence of water in different shapes (e.g. fountains) which promotes the vitality of the place
14. To have the possibility of doing actions – such as walking, watching, and the like – with a moderate or slow pace, promoting opportunities to take breaks in the space
15. To have the possibility of using the space in different weather conditions and seasons, contributing at the same time to its good state of maintenance
16. To preserve both the place's identity and the intangible characteristics of the site and its surroundings
17. To both allow and promote different types of functions such as games, breaks, walking and the like
18. To facilitate gymnastic activities with the presence of small equipment or a designated space
19. To have the possibility of doing actions that normally are not permitted – such as walking barefoot in the water or in designated public areas – improving a feeling of freedom and joy
20. To encourage the presence of art in its different forms
21. To promote sculptures, games or other elements and amenities which can bring a smile to a person's face favouring a state of liveability and happiness
22. To promote participation, namely the feeling of being able to contribute to the life of that place increasing the sense of belonging
23. To considerate the place as symbolic of the neighbourhood, improving the perception of its identity
24. To promote the educational function which a place has – for example, clearly displayed information about history of the place and the like or suitable ways to use it – increasing its intrinsic value.
25. To facilitate the use of new technology to increase the knowledge of its intangible values and history, offering a more profound experience of the place

## Phase 7: The design ideas from the users' points of view

Phase 7 concerns checking the emerging first design ideas – with the users of the place to obtain a "mosaic" of the degrees of pleasure of these.

Two kinds of information are collected. In the first kind of questionnaire, the design hypothesis is verified with demands selected by who carried out the study, suitable with respect both to the results of phase 6 and the object of the case study.

The second kind of questionnaire is constituted by the study of web reviews concerning booking of tourist services (e.g. Booking.com or TripAdvisor) or social network tools, including Instagram, Twitter, YouTube and Facebook. Information contained on these websites must be suitably analysed to obtain useful results. The number of followers, likes and reviews constitutes one of the main factors to consider.

Information by both the questionnaire on site and online will be overlapped, to comprehend the satisfaction on the current state and possible changes concerning the place, by common users of it.

## Phase 8: The project interventions

Following the check of consistency with the Charter in phase 6 and the results from phase 7, in phase 8 – the last of the method – the identification of the project interventions in the areas of consideration is carried out through the overlapping of the results of the previous phases and a check of consistency with spaces and urban furniture and equipment already present.

It is also important to connect the existent traces to the proposed project interventions that are not only the urban ones, but also the cultural and natural ones belonging to memories, perceptions and other tangible and intangible factors. A detailed analysis will provide more suitable information in order to make evident the existent characteristics and peculiarities and connect the project to new ones. In this way a map of urban health, liveability design will be obtained. The design of the new spaces to be harmonized with the existent ones is an important point above all when the final objective is constituted by the urban health, happiness and liveable achievement, concepts which are often hardly reachable.

# 12 Healthy Pl@ce app

The Healthy Pl@ce app is an application which can be realized to support the Healthy Pl@ce Design method to both easily view healthy place maps and related peculiarities detected with the surveys carried out as part of the investigation protocol and modify them with your own information and perceptions of the places in question (Sepe, 2010, 2012).

In the following, the general characteristics, functions and support in each phase of the method are reported.

The app is thought of for smart media, such as mobile phones and tablets as well as smartwatches, and is open source.

As for the main features, the Healthy Pl@ce App is (a) customizable, (b) interactive, (c) easy to use, (d) with intuitive graphics and (e) attractive. The app is customizable as it is possible – in editor functionality – to add or modify the elements on the map by inserting notes, images and comments that update the information contained in the map in the form of symbols. It is also possible to choose backgrounds and display modes for the symbol sheet to add further useful information to the place being analysed. Finally, it is possible to choose the smart media to use – PC, mobile phone, tablet, smartwatch – in order to get the best view.

It is interactive; namely, with the app, it is possible to download – always in open source – the base maps on which to insert the symbols; connect with other apps aimed, for example for calculating steps and times of running or other sports, natural, cultural or artistic routes; use a search engine to connect to social networks or other websites and download or upload materials; and look for information, such as services available (gyms or other), the degree or type of accessibility and different types of activities present in the place in question and times.

The app is intuitive and simple to use. It can be downloaded from specific websites or by itself on the site in question. The presence of a few simple buttons with an icon that recalls the functionality makes the app very easy to use as an editor and user, both in terms of maps and multimedia elements. The app is also designed for easy access so that it can be used by people with different abilities.

Finally, the app has attractive graphics in the various functionalities and uses, both for maps and for multimedia elements – photo, video and so

DOI: 10.4324/9781003098775-16

on – connected to symbols. The symbols are made with the use of bright colours that are in line with both the idea of ease of use and the concept of healthiness underlying the Healthy Pl@ce Design method.

The functions of the Healthy Pl@ce app include editor functionality, user functionality, user+ functionality and network functionality. The functions related to the analysis map and then to the project map are illustrated in the following discussion.

With the Editor function, it is possible to create the analysis map and the data cards connected to the elements present in it, modify maps and data cards and integrate these with materials on the network.

During the construction phase, the app loads the cartographic bases in open source from the internet and supports the insertion of the symbols related to the elements surveyed.

The symbols related to the map of health, and liveability analysis – in the fifth phase of the protocol – are pre-created with computer-aided design (CAD) software and imported from the app as a .jpeg or a .tiff.

The app also allows clicking on the symbol to open a related multimedia card, allowing the insertion of the main data related to it, including photos, information on equipment and street furniture present, activities carried out and relative frequency, questionnaires and related answers and websites.

The edit function can also be used on the individual cards linked to the symbol by adding further images, notes and answers to the questionnaires. In the editor function, it is also possible to calculate different types of indices (see the paragraph related to this topic in this chapter).

Finally, the app allows an easy connection to the main social networks, allowing photos, videos and other files useful for customizing the data cards related to the symbols to be downloaded.

With the user functionality, it is possible to directly view the map with the symbols and the cards connected to them without being able to modify them. With the user functionality, zooming the map is possible as well as consulting the cards by enlarging or shrinking some parts.

With the user+ functionality, compared to that of the simple user, it is possible to modify or add texts, drawings and images to maps and cards by using the bases already present and created by the Editor.

The network functionality allows viewing multiple health maps simultaneously in a network logic. This allows one to create itineraries and health paths between places near and geographically distant to visit, take advantage of their characteristics and healthy equipment and view and overlay multiple maps of the same place that have been edited by different people.

Regarding the functionalities of the project maps, with the editor functionality, the map is always created with the cartographic base used for the analysis map.

On the cartographic bases, the add symbol function and a pointer allow to position the symbols of the project interventions where detected.

The symbols related to the project map of health, map of health and liveability of the eighth phase of the method are pre-drawn with CAD software and imported from the app in a .jpeg or a .tiff.

By clicking on the symbol, the app also allows a related multimedia card to be opened, which allows photos, questionnaires and related answers and design references (with photos, sketches, drawings) to be inserted.

The app, as for the analysis map, allows, with the use of a drawing and text editor, the modification of the map information by creating any symbols or simple drawings and inserting notes of different types on the map. The edit function can also be used on individual cards linked to the symbol by adding further images, notes and answers to the questionnaires. With the app, it is also possible for the project map to calculate indices such as the Healthy, Liveability and Happiness Pl@ce Index, Creativity Place Index and the Environmental Activity Index using a preset formula. The calculation, which is explained in a dedicated section, is done by considering the results of consistency with the elements and factors contained in the map of health and liveability analysis with the principles of the Charter (phase 6 of the method). The calculation can be carried out using the planned project interventions of the map of the health and liveable design as well.

The app also allows an easy connection to the main social networks, allowing photos, videos and other files useful for customizing the cards to be downloaded.

Regarding the user function, as for the analysis maps, the map with symbols and cards is directly displayed. The user cannot edit them but can use the zoom function to zoom in and out on maps and cards.

Similarly, the user+ function, compared to the user function, allows texts, drawings and images to the maps and project cards to be modified or added. Since these are maps and project sheets, the changes can also concern general comments on the ideas proposed, or considerations in support of the ideas presented.

Finally, the network function allows to simultaneously view multiple health project maps created by people of different backgrounds in order to connect itineraries and routes.

Accordingly, to face both the overlapping of the multimedia elements collected by the method and the new needs of soft technology expressed by people, the app could be connected to QR codes posted on-site from which both further information about the site, itineraries and places derived by the maps which are carried out can be downloaded. The main users to who the app is devoted include visitors, tourists, technicians and administrators who are interested in more healthy, liveable and sustainable places.

As specifically concern the support of the app to each phase of the method, in the first one, namely that of the identification of the area, the app allows the visualization and delimitation of the place in the first window. In this way, in other windows, the selected parts of the area can be visualized. A zoom function allows the area to be scaled, depending on the necessity of use.

In the second phase, concerning the identification of the activities, perceptions and contributing elements, the app has a function for importing the collected information; schematizing on the map the movements of people, urban furniture and equipment and the activities carried out; and calculating the percentages of these items. Furthermore, the visualization of elements and movements is supported by images and video.

In the third phase, concerning questionnaires, the software imports both the questions administered during the site inspections and the answers received. The results of these answers are reported. The app will visualize the percentages of answers given, allowing from the map the comprehension of the factors of major interest concerning urban liveability and healthiness which emerge from different types of users.

In the fourth phase, the traditional analysis of the places in object is imported and visualized in multimedia way with pictures, both new and historical. The map can be downloaded in open source, and this can be supported by news present in the web devoted to the knowledge of both the current state of fact and the history of the object area.

In the fifth phase, the map of health, liveable and happy place is realized by inserting the relative symbols imported in a .jpeg or .tiff. The symbols are related through the app to a pre-organized database format which may contain many kinds of information collected during the surveys. The app allows multimedia information – including photos and videos – to be uploaded as well as the internet to be searched.

In phase 6, the principles of the Charter of Urban Health and Liveability Design, which are imported before the beginning of the first phase actions, will be indicated on the previous map to visualize the consistency between the elements and factors on the map and them.

In phase 7, the software imports both the questions administered during the sight inspections and information by the social networks. The results of these two kinds of answers are visualized on the map through the percentages of answers, allowing different types of users to comment on the proposed project interventions to be comprehended.

In phase 8, the map with the project interventions synthetized in symbols is created, and each symbol is associated to a pre-organized database which can contain multimedia data.

# 13 Indices

The type of data collected with the Healthy Pl@ce Method and the synthesis represented in the maps offer the possibility of calculating different indices (Sepe, 2019b).

The main index concerns the Healthy, Liveability and Happiness (HLH) Pl@ce Index, which collects and reduces the degree of healthiness of the place to a number. More specific indices can then be created, from information on particular data such as the Creativity Place Index, which summarizes the data relating to the presence of artistic elements capable of giving a perception of liveability to the place, and the Environmental Activity Index, which expresses with a number the presence of natural activities that involve people offering a perception of healthiness to the place.

The HLH Pl@ce Index is defined by the degree of healthiness, liveability and happiness of a place and its activities according with the results of consistency with the Charter (phase 6 of the method).

In particular, this is calculated by the sum of all factors deriving by the comparison with the principles of the Charter associated to a weight. The weight is calculated between 1 and 5, namely, wi = 1 has little impact in calculating the index, while the highest value, wi = 5, counts for a lot.

For example, regarding principle 2, "To encourage the use of the place by people of different age groups, from children to older adults", the weight is 1 if in the place in object few activities for different uses and for different people groups are previewed and 5 if the place is well equipped for all.

The index can be calculated in different periods of the year in order to measure the impact of the activities in different weather conditions. The range is from 25 (minimum) to 125 (maximum).

Accordingly, and with the same range, other indices can be calculated, including the Creativity Place Index and the Environmental Activities Index.

The Creativity Place index is calculating summing factors deriving by the comparison with some principles of the Charter associated to a weight.

In particular, these concern the following:

3. To eliminate architectural barriers which might discourage people from frequenting that space
6. To retain an adequate state of cleanliness and maintenance

DOI: 10.4324/9781003098775-17

12. To have direct contact with natural materials, preferably local, used in the design of the space
20. To encourage the presence of art in its different forms
21. To promote sculptures, games, or other elements and amenities which can bring a smile to a person's face promotes a state of liveability and happiness

This index is obtained by adding and giving a weight from 1 to 5 to for example the artistic playgrounds or furniture with a particular design, sculptures depicting subjects that involve people, historical or design bridges and scenic or engaging fountains associated with them. The range is between 25 and 125. For example, an artistic sculpture, depending on the state of maintenance and the aggregation of people it creates, has a weight of 1 to 3 while a sculpture with subjects that create a lot of interaction, joy and a feeling of well-being between people can have a weight of 4 to 5. To list other examples, there are playgrounds that have artistic forms that have a good weight in terms of cheerfulness and joy, but a playground that in addition to having artistic forms is also built with natural materials has more weight. Weight 1, the lowest, is given to artistic elements in a poor state of maintenance and positioned in a part of the area that is difficult to access. Weight 5 is given to an artistic element that creates a lot of involvement, and/or is made from natural materials. It also depends on the context; for example, in a place that is affected by severe climates for most of the year or that is lightly inhabited, the involvement of people will certainly be less than in a place where the climate is generally mild and/or the area is very populated.

With the same criteria, other indices can be calculated, including the Environmental Activity Index to calculate the influence of natural activities on the perception of liveability and healthiness of people.

The reference principles for the calculation include the following:

4. To create a suitable balance between the elements of nature, landscape and equipment in the composition elements of the space
6. To retain an adequate state of cleanliness and maintenance
11. To fully perceive naturally occurring smells – for example, wood, grass, sea
12. To have direct contact with natural materials, preferably local, used in the design of the space
13. To have the presence of water in different shapes (e.g. fountains) which promotes the vitality of the place.

In this case the presence of flower beds and trees have a weight that can vary from 2 to 3 according to the quantity and position, while nature trails, water sports and picnic areas on the lawn will have a weight from 4 to 5 for the involvement of people.

The presence of a fountain or a body of water can have a weight of 1 to 5 depending on the interaction of people, direct contact with water and the pleasant perceptions it creates. A weight of 1 is given for example in case of bad management of greenery or other natural elements, while a weight of 5 is given in the case of a good state of care of the natural environment and the perception of joy that the body of water creates.

Finally, the three indices can be also calculated using the project map and the correspondence of the planned project interventions and the principles of the Charter, in order to understand how much the indices would change compared to the current situation.

# 14 Observation

Here, health is meant as factor important not only to resolving diseases but also to improving well-being, which can be supported by a suitable way to design places and public spaces, as illustrated in the theoretical section of this volume. Liveability is a factor that is also important in urban design and is related to a more general perception of beauty, quality and comfort of the place, not necessarily in connection with health. Urban happiness can be given by both perceptions and intangible aspects and is strongly connected to architectures, public spaces and the natural environment, although with a certain degree of subjectivity. Indeed, urban health, liveability and happiness are three concepts that are strongly interwoven and are related to the sustainability and well-being of both people and places. Identifying the presence of these three factors within a site and realizing an urban project which contain these is a great challenge. The Healthy Pl@ce Design method is in this sense very useful.

It has been experimented with in different contexts in Europe, Canada and China. The experiments concerned public spaces which are well known and successful in order to comprehend the behaviour and movements of people and the physical elements which contribute to urban health, liveability and happiness. This is carried out through a protocol with different kinds of surveys; among these is the survey realized using *ad hoc* data cards.

A focal point is the identification of the intrinsic and extrinsic elements which contribute to urban liveability, health and happiness. Indeed, those elements are in strong relation each other, and consequently, the identification of the areas which need to be improved from those points of view is complex.

The protocol of investigation was tested in its single phase and then in its whole. The main observations deriving from the different tests will be reported in the following.

In the first phase, the choice of the area to analyse is given by site inspections of the place at different hours and days to decide the parts of an area characterized by the continuity of paths, activities, equipment, urban furniture and green space. Often the borders of the site to analyse are already given by the area in itself – for example as for Madrid Rio, Promenade du Paillon and so on – and it is easier to decide the perimeters. At other times,

DOI: 10.4324/9781003098775-18

there are more open spaces within a wider area, and it is more difficult to determine the continuity because different elements and activities are present. The overlapping of information given by observations on-site and the observations of the planimetry of the place will determine the more suitable borders of the place to be analysed. These could be in any case modified if, during the next phases, new, useful data to determine the borders are observed.

In the second phase, it is important to use a database for the surveys, but at the same time, it needs to be observed whether further elements emerged during the site inspection that need to be analysed. The places are different each other and different categories of elements can emerge which were not previously considered. In particular, scenic fountains, artistic playgrounds or specific paths are some of the elements that could change from one place to another.

The information collected in the third phase are those carried out by the users of the place in question through a questionnaire. The demands of the questionnaire can be changed according to the peculiarities of the area, substituting some or adding others. The questions should be open, leaving people free to say what they think and perceive.

The fourth phase is the more objective of the method, although it is not so easy to traditionally analyse what is really useful for the topics of urban health, liveability and happiness. Furthermore, also the urban design or planning plan could be of particular importance for comprehending the present and the future of the place, possible trends and the typology of the place which could be realized. Also, in this case, it needs to suitably overlap the information collected to understand details on the elements which constitute and contribute to the health, liveability and happiness of the place.

The fifth phase is the more important part of the analysis: it needs to synthetize all the elements collected during the previous phases; during this phase, symbols need to be created that are not only friendly, easy to draw and easy to comprehend but also flexible to allow inserting new symbols. The quantity of symbols and information on the map should be suitable to allow the map to be easily read. Another point is the legend, which should show what is contained in the map. It is important that the legend is clear but not too long to allow for quick comprehension.

In the sixth phase, the consistency between elements on the map of analysis and the principles of the Charter of Urban Health and Liveability Design needs to be comprehended. The consistency can be observed between only few principles or more, or, again, the case could show further factors which could lead the addition of new principles. Furthermore, data to calculate indices could be collected.

Regarding the seventh phase, it is similar to the third. The difference is that the questionnair from phase 3 serve to understand how people perceive the health, liveability and happiness of the place in question; in phase 7, these serve to understand what people think about the project interventions which were identified. Furthermore, there is information concerning the project interventions collected through the social networks. It is not easy to interpret the comments in a manner which is useful to the case study objective; if these are not suitable to use regarding the objective, only the general information concerning hashtags, followers and so on are reported.

In the eighth phase, the map that must be carried out has to show the project interventions that have to be carried out to enhance the health of a place. The project interventions are the final synthesis of all the previous phases of the method and, for this reason, the most important. The problem that sometimes happens is that these are too general; however, the more detailed the description of the project interventions, the easier the translation into design ideas will be.

In any case, it needs to take into account that the design hypothesis reached following the method is – given some necessary elements – only one of the possible solutions. From Healthy Pl@ce Design method and according to the health and liveability principles, many design ideas can be derived and translated into drawings in different manners.

The app is a support for the dynamic visualization of the results of the maps; this is useful for its flexibility and different functionalities and is customizable for many kinds of users.

Finally, the indices can be useful in this sense for measuring both the results of the analysis and those of the design. The indices indicated are only some that can be calculated; from the information gathered, many indices can be created and be useful for different aims.

# Part IV

# Case studies

# 15 Bordeaux

## The area

Bordeaux is a city in New Aquitaine in France. The visits which were carried out on-site had the objective of deciding the parts to be included in the study area. Observations were carried out in the morning and afternoon hours during the week and weekend.

The area that has been decided to be analysed is the left part of the Garonna waterfront, which starts at Pont Saint Jean and ends at Pont Jacques Chaban Delmas. In order to make all the surveys, analysis and questionnaires, the area was divided in four stretches, which include the waterfront in itself with, in some cases, internal places, which are characterized by urban continuity with the waterfront.

The four sketches are Pont Saint Jean (Quai Sainte-Croix)–Pont de Pierre (Quai des Salinières – Quai de sports Saint Michel), Quai Richelieu–Quai du Maréchal Lyautey, Quai Louis XVIII and Quai des Chartrons–Quai de Bacalan–Pont Jacques Chaban Delmas.

The whole study area is characterized by a cycle/pedestrian lane which borders the external part of the place in question, light poles, benches and trashcans with a unique and beautiful design and, on the river, railing with the indication of the length of the waterfront.

## People and activities

The analysis begins in the first stretch, namely Pont Saint Jean (Quai Sainte-Croix)–Pont de Pierre (Quai de sports Saint Michel–Quai des Salinières).

In the first survey of the second phase, the types of people who use this place are observed. This stretch is mainly used by locals of all ages and children. There are also workers, including gardeners, and people on their way to work (e.g. at the Conservatory of Music which is nearby).

The second survey concerns the activities carried out and relative attendance. People, including professionals, use the cycle/pedestrian path by foot or by bicycle with average frequency. Young adults and adults also perform running activities with medium frequency. The locals carry out

DOI: 10.4324/9781003098775-20

break and/or observation activities both on the benches and on the lawn with a medium-low frequency. Children and young people, as well as adults, especially locals, play various sports activities in the Parc de Sports Saint Michel, including pelota, gymnastics, hockey, basketball, urban football and beach volleyball. Teachers and animators frequent these spaces as well. These activities are carried out by all people with medium-high frequency. In the summer months, in this part of park, people listen to music and dance with high frequency.

The third survey concerns the single and mixed perceptions. The pleasant visual perceptions concern the Pont Saint Jean and Pont de Pierre bridges that can be seen from Quai Sainte-Croix, the Garonne River, the right magazine of the river and ancient buildings on it, the historic buildings and the Burgundy Gate that can be seen on left of the cycle/pedestrian path on Quai de la Monnaie and Quai de la Grave. Finally, there is a pleasant perception of both the row of trees and the hedges along the path.

Acoustic perceptions mainly concern that of the Garonne on the one side, which is pleasant, and of vehicular traffic on Quai de la Monnaie and Quai de la Grave, on the other, which is annoying. In the sports areas, the noises related to sports activities and therefore to the movement of people who practise that specific activity or the equipment used to play are perceived – all pleasant. Tactile perceptions mainly concern the lawn and sand in the areas of the Parc de Sports Saint Michel, all of which are pleasant. The olfactory perceptions are pleasant and concern the lawn, the flowers and the vegetation very present in the various parts of this stretch, as in the following ones. There are no particular taste perceptions.

Mixed perceptions related to this stretch include dynamism in sports areas, especially when these are used by several people: calm and serenity in the parts of the internal promenade that runs along the river and calm mixed with chaos during the hours of increased vehicular traffic on Quai de la Monnaie and Quai de la Grave.

The equipment includes a cycle/pedestrian path, at the outer edge of the area, a pedestrian path at the inner edge that runs along the Garonne, trees and greenery, benches and street furniture such as rubbish bins, lighting poles, poles for bicycle parking, the sport equipment and concrete, green and sand fields created for the sports activities.

The second section concerns Quai Richelieu–Quai du Maréchal Lyautey. The people who use this stretch are many: locals, residents, tourists and professionals of all ages. Being one of the tourist parts of the city, but at the same time having work offices on the other side of the park, the percentage of the type of people varies both during the day and in general within the area. In the Quai du Maréchal Lyautey section for example, there is a greater presence of tourists than in the Quai Richelieu section.

With respect to the activities carried out, the people who use this stretch are professionals who go to work along the cycle/pedestrian path by foot or by bicycle with average frequency. Young people and adults perform running and walking activities also use this stretch with medium frequency. The locals carry out break activities, picnics on the grass, observation of the panorama of the river and the flowery meadow both on the benches and on the lawn with a medium-low frequency. The activities carried out by locals and visitors of all ages with medium-high frequency and for long times concern observing the Water Mirror in Place de la Bourse and the activities people carry out crossing it; in particular, children and teenagers play with the water and the steam of the Water Mirror, often for a prolonged duration.

Regarding the perception surveys, the single perceptions include the visual of Water Mirror of Place de la Bourse, the gardens and the river bridges; the smell of flowers; the sound of people playing in the Water Mirror; and the tactile of people who sit on the gardens. The mixed perceptions concern the joy, happiness, serenity, healthiness and the liveability who people feel using this place. With respect on the furniture, the main concern benches and light poles.

The third stretch is Quai Louis XVIII and the first survey concerns the type of people who frequent this place. This stretch is widely used by locals, residents and tourists due to the presence of a large green area that continues in the Parc des Allées de Chartres on the other side of the river and ancient sailing ships; the latter is used heavily during the wine festival held annually in June. The large stretch of riverside allows people of all ages to use it. In addition, the place is also very popular with workers who carry out various activities on the riverside. The activities are carried out in this stretch above all by locals and visitors – both young people and adults – with a medium-high frequency and for a long time. These include cycling, lying on the lawn and under the trees of the park, strolling along the seafront, strolling on the Parc des Allées de Chartres, sitting on the wooden platforms, taking a break, having a picnic, asking for information and taking a boat (cruise boat, etc.), watching and visiting ancient boats, going to the market for typical food products and sitting at the tables in the bar or restaurant (Figures 15.1–15.7).

Furthermore, locals – both young people and adults – run, observe the area and sit on benches with a medium frequency and for medium time.

The equipment in this stretch includes the cycle and pedestrian path, a green park, wooden platforms for sitting, the cruise terminal, cafes, sailing ships, benches, bollards and lighting poles with a pleasant design.

With respect to the single perceptions, all pleasant, these include the visual perception of the cathedral, the cruise ships, the park, the statues and the skyline of the other riverside; the tactile perception of the meadow and the wood of the benches; the smell of the green, the plane trees and the water of the river; the sound perception of the passages of people on bicycles, the cruise departure and the voices of people at the market; and the taste perception of the local and bioproducts. With respect to the mixed perceptions, these include happiness and liveliness near the sailing ships and serenity, joy and health along the paths on the green.

The fourth stretch is Quai des Chartrons–Quai de Bacalan–Pont Jacques Chaban Delmas. The stretch is widely used by locals, residents, professionals

*Figure 15.1* Bordeaux riverfront, detail of a playground for children

*Figure 15.2* Bordeaux, riverfront, the Water Mirror and people playing with the water vapour

*Figure 15.3* Bordeaux, riverfront, the gardens in front of the Water Mirror

*Figure 15.4* Bordeaux, riverfront, the old and new boats

*Figure 15.5* Bordeaux, riverfront, the market with local products

*Figure 15.6* Bordeaux, riverfront, skate park

*Figure 15.7* Bordeaux, riverfront, shops and restaurants

and visitors of all ages. Due to the main activities and attractions, this part of the area is used by visitors and workers, young people, adults and older adults.

The activities include walking, strolling, running, cycling, going on the skateboard, sitting on the grass, having a picnic on the grass, observing the river panorama, observing people skateboarding in the dedicated park, observing the people walking by, sitting, taking a break, going to work, buying organic products at the market, going to the bar or restaurant, shopping and going to the science museum.

As in the other stretches, the equipment includes urban furniture with benches, rubbish bins, lighting poles with a unique and pleasant design and a wood platform for sitting and bollards. Furthermore, there is a local and bio-open-air market, the Colbert skate park, pubs, restaurants, shops and the science museum.

Single perceptions include pleasant visual perceptions of the Citè du Vin, Saint-Louis Church of the Chartrons, CAPC Contemporary Art Museum, the Jacques-Chaban-Delmas bridge towers, the cross of tram, and the skatepark murals; the taste and smell perceptions of the market of bio- and local products; the smell perceptions of the river and green area; and the acoustic perceptions of people voice, bicycles, the tram and cars.

The mixed perceptions concern dynamism in the parts closer to the skate area, and joy, happiness, health, comfort, old–new mix in the other parts – all pleasant.

## The places from peoples' points of view

After the surveys, 130 people – residents aged between 20 and 65 years – and visitors from all over the world – mainly from Europe and the US, aged between 30 and 60 years – were administered a questionnaire. The languages used were English and French. People were generally willing to give answers to the questions.

The questions included the following:

1. Does this place give you a feeling of happiness or sadness/liveability or discomfort/healthiness or unhealthiness?
2. What are the elements that give you the above sensations?
3. What are the main facilities that give quality to this place?
4. What are the activities that you do in this place and how often?
5. What do you think about the presence of many or few people here? Do you think that it is capable to improve the pleasantness or unpleasantness of the place?
6. According to the place healthy, liveability and happiness what could be done in order to improve this place?
7. What is a healthy/happy/liveable place that you remember in this city or elsewhere?
8. How might the weather conditions influence the perception of healthiness, liveability or happiness of this place?

In the following, questions and answers are reported.

To the question, "Does this place gives you a feeling of happiness or sadness/liveability or discomfort/healthiness or unhealthiness?" all people answered – independent of nationality and age – that the place gives a feeling of liveability and health and, in the Water Mirror stretch, a perception of happiness.

To the question, "What are the elements that give you the above sensations?" although everybody answered the overall presence of well-designed spaces, the general atmosphere and high presence of green area, people observed different elements also in relation to the part of the area in which the question was administered. Sixty per cent of locals and visitors answered the comfort of the walk, namely the orientation that is clear, and the pavement that is of good quality and constant level, allowing people to walk and enjoy the place or take a break on the meadows. Forty per cent answered the presence of the river and the Water Mirror, which widen the beauty of the place and the historic elements, such as the bridges and the old buildings.

To question 3 – "What are the main facilities that give quality to this place?" – people answered in different ways depending on age. Young people answered the presence of the bike lane; the Parc de Sports for pelota, gymnastics, hockey, basketball, urban football and beach volleyball; the skatepark; and the green area in general to take a break. Of adults, 60% answered the many benches and the different areas to sit, the bike and pedestrian lanes and spaces, the many places for babies and the Water Mirror. The locals added that the Water Mirror is useful in the warmy periods and beautiful to see in the other periods of the year. The other 40% of adults answered the shops, the bars and restaurants, the boat terminal and, near, museums and historic centre. Older adults answered the presence of benches close to beautiful places to see, such as the river, the green areas, the historic buildings and, in general, the beauty and easy walk.

To the question, "What are the activities that you do in this place, and how often?" 60% of locals and visitors answered walking, taking a break, cycling, observing, buying local food products, staying or crossing the water mirror and taking the kids for a walk. The other 40% of locals and visitors answered running, eating or drinking, shopping and taking a boat or visiting a traditional sailing ship. Of these 40%, all the young adults added playing sports and games of different kinds, staying on the meadows and meeting friends.

To question 5 – "What do you think about the presence of many or few people here? Do you think that it is capable to improve the pleasantness or unpleasantness of the place?" – about 100% of those interviewed answered that the presence of many people improves the pleasantness of the place. The place has a great extension, and there are many activities to do; giving a sensation of comfort also if there are many persons.

To the question, "What could be done in order to improve this place according to the place healthy, liveability and happiness", visitors answered nothing, adding that the area transmits good feelings which improve health and liveability aspects. Visitors added that the feelings of happiness are perceived close to the Water Mirror. Locals, however, answered that the general atmosphere is good, with about 60% of adults answering that other playgrounds for children could be localized in other parts of the riverfront. The remaining 10% answered with different ideas, including the presence of more green areas to decrease the noise of the vehicles in some parts of the place. Of this 40%, the 10% of the young adults answered that they would like to have more space for art performances.

To the question, "What is a healthy/happy/liveable place that you remember in this city or elsewhere?" local people mainly answered that this is a unique place; 10% of them added that a similar kind of feeling can be found on the French Blue Coast. Visitors answered many kinds of places, including the banks of the Seine in Paris, Venice Beach in Los Angeles, Thessaloniki (Greece) and Alicante (Spain) new waterfronts.

Finally, to the question, "How the weather condition might influence the perception of liveability or happiness of this place?" 80% of locals answered that this place is used both in warm and cold periods, while 20% of them answered that in very hot periods, the presence of the Water Mirror is refreshing but that in the rainy periods, it is not used very often. All visitors answered that this is a good place in all the seasons, because of the presence of the Water Mirror and of bars and restaurants.

## The transformation of the area

The traditional analysis gave the result that follows. The urban regeneration in Bordeaux is based on an articulated planning of urban interventions and financial resources aimed at redefining the identity of the places, which has long been linked only to wine production, and revitalizing the economy of the city, placing it among the top French metropolises (Clout, 1984; Lydie, 2003; Tsiomis and Ziegler, 2007).

Indeed, Bordeaux is a medium-sized city in France, the capital of the Gironde Department and the Aquitaine region, with an economy traditionally characterized by the wine sector and tourism related to it.

In the 1940s, the river port, located 50 kilometres from the sea, was a commercial port. Today, commercial traffic has been diverted to Le Verdon, and the hangars have been dismantled to create a series of public spaces on the waterfront and enhance the view of the buildings from the quay, while the port has been transformed into a cruise landing in continuity with the tourism development of the city.

The main project is "Bordeaux les deux rives", aimed at the integration, development and revitalization of the territory and landscape of the two banks of the Garonne River (Martone and Sepe, 2012).

The first phase for the implementation of the Bordeaux transformation strategy is organized following some main actions, the realization of the mobility project, in particular linked to the tram network, the redevelopment of public spaces, especially in relation to the quais jardinés.

The urban regeneration process began in 1995 when Alain Juppé was elected mayor of Bordeaux and later president of the Bordeaux Urban Community (CUB), which administers the metropolitan area and promotes the construction of a new urban project and a tram network to support the transformation in an integrated way.

In 1996, the PPU (Urban Pilot Project) "Bordeaux les deux rives" was presented in response to the call for tenders launched by the European Commission under the European Regional Development Fund funds. Through the Bordeaux les deux rives, the development of the two banks of the Garonne is promoted by translating the city's recovery project into suitable development and revitalization strategies for the territory connected to its river.

The program was implemented between 1997 and 2001 with the coordination of the public–private company Bordeaux Métropole Aménagement (https://www.b-m-a.fr), of which the CUB is the main public shareholder together with the cities of Bordeaux, Mérignac and Arcachon and the Conseil Général de Gironde, while Caissedes Dépôts et Consignations, Crédit Local de France, Caisse d'Epargne Aquitaine Nord and Crédit Agricole are the main private shareholders. In this regard, 22 actions were organized for the realization of an integrated project for urban requalification, economic development and social and cultural revitalization. The actions concern the tangible and intangible cultural heritage and include interventions, such as the redevelopment of the system of open spaces, the reunification of the historic and contemporary city with the river area, the Plan lumière, the restoration of monuments and historic facades and interventions on the building heritage.

Particular attention is paid to the environmental rehabilitation of the banks of the Garonne, to the creation of a sustainable mobility network to obtain the most adequate accessibility to the different parts of the territory affected by the transformation, to the creation or enhancement of activities to ensure a high level of tourist reception throughout the year with cultural events linked to places – such as the Fête du Vin and Fête du Fleuve – and to the participation of the population. The program also included the construction of equipment and accommodation for a new residential centre located in the Bastide area, and a large area of public facilities, including the botanical garden, sports, play and free-time areas (Carta, 2007).

Regarding mobility, new axes and development nodes were identified to re-organize pedestrian traffic, bicycle and car paths and an extensive tram system (Figures 15.8 and 15.9).

Of the new tram lines, designed by an interdisciplinary group of architects and landscape architects, one runs parallel to the system of external boulevards and the other along the Garonne, serving the quays for a total of 23 kilometres. The perception of public spaces, due to the redevelopment of the surrounding urban context, was significantly changed.

*Figure 15.8* Bordeaux, images from the city centre with the new tram network

*Figure 15.9* Bordeaux, detail of a renovated square in the historical centre

For the projects of the open spaces, an important reference is constituted by the Charte des Paysages of Bordeaux, an essential document of the Plan Vert of the city approved in 2000, in the drafting of which the landscape architect Michel Desvigne collaborated.

The spaces are created by reinterpreting the Bordeaux tradition – through the choice of materials, furnishings, botanical species, themes and contemporary art objects in line with the identity of the specific place – or by re-creating a different one but in harmony with the characteristics of the context. The fundamental thematic axis for the construction of Bordeaux les deux rives is the system of historic open spaces on the left bank which were involved in a total redesign carried out by Corajoud.

Among other projects within the PPU Bordeaux les deux rives, the ZAC des Chartrons project implemented the redevelopment of an abandoned industrial area in a state of decay and allowed the creation of a new neighbourhood consisting of houses and commercial buildings, having an effect on the whole area between the historic centre and the submarine naval base north of the city.

In 2001 the Schéma directeur de l'aire métropolitaine bordelaise was approved for the 2010–2020 period. The Schéma Directeur which has value as SCoT (schéma de cohérence territoriale) defines the strategic planning and sustainable development of an area that covers 91 municipalities and 820,000 inhabitants by 2020. To this end, the objectives of the schéma concern limiting urban expansion; focusing development on the heart of the city and peripheral centralities; preserving, enhancing and managing the natural structure of the landscape surrounding the city; defining and structuring the major economic centres of city development; and promoting urban regeneration and densification along heavy transport axes.

The last part of the analysis concerns the identification of places and activities which make the place liveable, healthy and happy from the users' point of view.

## The liveability and healthiness factors

The new waterfront of Bordeaux is constituted by an area which is extended for about 4.5 kilometres from Pont Saint Jean to Pont Jacques Chaban Delmas.

The area is widely used by locals and tourists as it is easily accessible both with respect to the city centre and its less central areas and in terms of entrances that can be used by people of all ages and motor skills.

The waterfront runs along the Garonne River on one side, while, on the other, along a cycle/pedestrian path, in turn, is flanked by a wide road for cars, buses and trams.

The place is generally characterized by a large presence of green in its various forms that acts as a connector for the different areas that characterize this part of Bordeaux: trees, lawns, gardens, parks and hedges. And every form of green area has both an aesthetic function and a connection and is a multiplier of healthiness and liveability. The quais jardinés of Place de la Bourse with their flowery lines almost transform into a sort of impressionist painting when people sit there, who unknowingly become an integral part of the urban scene, allowing this place to actually become a place for people.

The promenade lawn on the Quai Saint Louis faces the parks on the Place de Quinconces and the Parc des Allées de Chartres and the Jardin Public backwaters. Throughout the waterfront, on the other hand, trees and hedges draw the cycle/pedestrian and mixed path, offering a perception of liveability and healthiness.

They are flanked by lighting made with scenic lampposts placed all along the street and, in the section of the water mirror, by lights that enhance both the Place de la Bourse and the Corajoud's work, making this place pleasant and liveable both day and night.

Each section of the riverfront is characterized by different functions: play and sport with fields and equipment of many types; market of typical products; pause, rest and observation; walking, cycling and running; skateboarding workouts; crossing and refreshments in the water, picnics and visits to historic sailing ships; refreshments in bars and restaurants; shopping in the shops of the Quai des Chartronne; and visiting the Cap Sciences museum. These activities are all carried out with a medium-high frequency, in particular in relation to the Water Mirror, which offers a particular feeling of liveability, healthiness and happiness to those who visit it. The most attractive part of this place is, in fact, the Water Mirror not only for its beauty and scenery but also for allowing people of all ages to look at it, cross it, play with water and cool off in summer.

The large presence of both visual and tactile perception made by natural elements, such as the river, the lawn and the park with trees and flowers and the sand, create feelings of tranquillity, liveability and healthiness.

The presence of ancient sailing ships and cruises, the buildings in Place de la Bourse, the Meca, Maison De L'économie Créative et Culturelle, the Cité du Vin create a particular mix of architecture that gives joy to the people who visit the area (Figure 15.10).

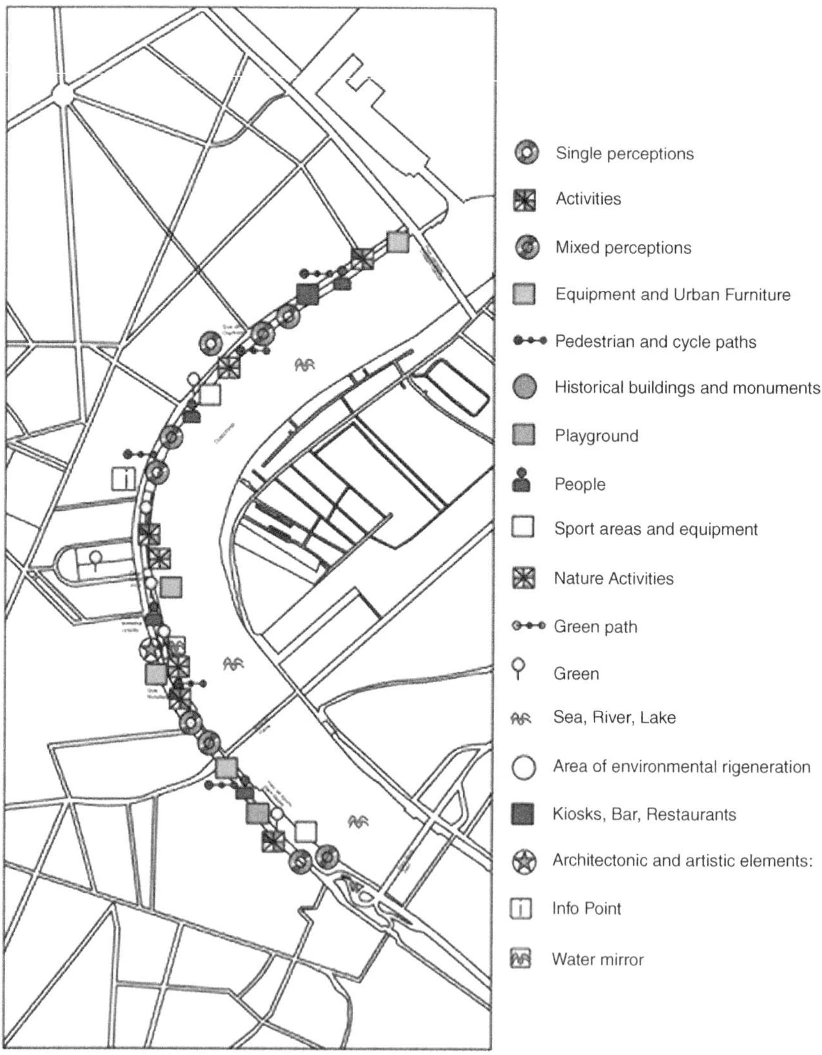


*Figure 15.10* Bordeaux riverfront, the map of urban health and liveability analysis

## Designing health and liveability

The check of the degree of health, liveability and happiness is carried out through the 25 principles of urban health and liveable design concerning the Charter. By the check of the principles, the Garonne left side area is in line with almost all. In all its extension, as it emerged from the previous phases, the path contains a wide variety of elements which makes it healthy, liveable and happy.

The principles which could be improved include (9) to minimize or eliminate the noise generated by public transport; (15) to have the possibility of using the space in different weather conditions and seasons, contributing at the same time to its good state of maintenance; (20) to encourage the presence of art in its different forms; (21) to promote sculptures, games or other elements and amenities which can bring a smile to a person's face creating a feeling of liveability and happiness; (24) to promote the educational function which a place has, increasing its intrinsic value; and (25) to facilitate the use of new technology to increase the knowledge of its intangible values and history, offering a more profound experience of the place.

The items of the questionnaire, which follows, derive from the principles that could be better satisfied and elements collected in the previous phases:

1. Do you use this area during all periods the year? What do you think to create urban public light structures to use this place also in cold periods?
2. What about inserting old or contemporary art products in its different forms to contribute to the sense of liveability and happiness of this place?
3. What do you think about widening the game areas inserting more playgrounds and places for games for people with disabilities and older adults?
4. What do you think about inserting more information points (such as totems, maps and so on) with the process of regeneration of this place?
5. What about widening the use of new technologies to give a more profound experience of the place (e.g. virtual maps, virtual experience connected to the history and local products such as the wine, etc.)?
6. What about minimizing the sound of vehicles in the parts where this is more evident?
7. What about creating an area devoted to dogs?

The questionnaire was administered to about 140 people – aged between 20 and 65 years – locals and visitors from all over the world – mainly from Europe and the US. The languages used were English and French, and people were interested in giving answers to the questions.

In the following, the questions and answers are reported. With respect to the first question, "Do you use this area during all periods of the year? What do you think about creating urban public light structures to use this place also in cold periods?" 100% of locals answered that they agreed because they often used this place but that there were some cold and rainy periods in which they did not use it a lot. Regarding tourists, 80% agreed with this idea, adding that they visited this place above all in the summer period. But if there were light structures for using the waterfront during cold periods, they will visit it also in those days. The other 20% answered that in the cold periods, they would prefer closed places.

To the question, "What about inserting old or contemporary art products in its different forms to contribute to the sense of liveability and happiness of this place?" 90% of both locals and visitors aged between 30 and 60 agreed with this idea. The other 10% of interviewees thought that artwork could be added only if related to nature, namely flower sculptures or water fountains or the like. Of locals and visitors aged between 20 and 30, 80% agreed adding the interest for a kind of art such as murals – close to the skate or sport areas – or participative sculptures. The other 20% answered that the area was already pleasant in this way.

With respect to the question "What do you think about widening the game areas inserting more playgrounds and places for games for people with disabilities and elder people?" the answers to this idea were generally the same. Independent of age, 100% of locals answered that they agreed to add or improve game areas for all. Of visitors aged between 20 and 30, 100% agreed with this idea, adding to include areas for street sports, parkour and street workouts.

Regarding the question, "What do you think about inserting more information points (such as totem, maps and so on) with the process of regeneration of this place and information?" the answers to this question were similar to the previous question. One hundred per cent of both locals and tourists, independent of the age, answered that they agreed. Many young people added that they would prefer interactive maps, QR codes and social networks with information concerning the regeneration as well as information about activities and equipment, anticipating in this way the next question.

With respect to the question, "What about widening the use of new technologies to give a more profound experience of the place (e.g. virtual maps, virtual experience connected to the history and local products such as the wine, etc.)?" 100% of both locals and tourists aged between 30 and 65 agreed with this idea. The tradition of Bordeaux is related to the wine and many well-known local products, so they agreed to the idea to know more about it and where to buy it. Young visitors agreed, adding their interest to have the new technologies also related to social networks.

To the question, "What about minimizing the sound of vehicles in the parts where this is more evident?" the interviewees were generally agreed. One hundred per cent of locals and visitors noted that in some parts, the car lanes are quite close to the path. They then agreed about minimizing the sound of vehicles in these parts, adding that in other parts – those related to the Water Mirror and the gardens – this is not necessary.

Regarding the question, "What about creating an area devoted to dogs?" 100% of interviewees agreed with this idea. Although not all the interviewees had a dog, they agreed that dogs should have their own area to play and stay with other dogs.

Checking social networks and websites with suggestions, no official Instagram page is present, but many pages exist with the #Bordeaux hashtag, such as #bordeauxmaville, which has about 905 posts and 115,000 followers. There is no Facebook page, but there are tourism pages, such as Visit Bordeaux, with about 137,687 followers, and @Bordeaux on Twitter, with 230,898 followers, and many movies on YouTube. Among the travel websites with suggestions, on TripAdvisor, there are the presence of Place de la Bourse, with about 7,000 comments; the Water Mirror, with 6,500 comments; La Cité du Vin, with 5,660 comments; the Marche des Quais on Quai de Chartrons, with about 500 comments; and Pont de Pierre, with about 1,350 comments.

The comments are all positive and emphasize in particular the beauty of the Water Mirror, the particularity of the Sunday market and the architecture and the interest of the Cité du Vin.

From the information gathered in the previous phases of the project, the following project interventions resulted.

The first project intervention is to create urban public light structures so this place could also be used during cold periods. Bordeaux is a city with a mild climate, but there are very rainy winter periods. The possibility of being able to use this place even in winter would be useful both for the inhabitants who could use this area all year-round and for tourist offerings of the city. It would involve creating semi-closed and light structures with a design in line with the identity of the place. The most suitable parts for the positioning of these structures are those where there is a lower visual relationship with the river in order to minimize the impact on the environment.

The second project intervention is to insert old or contemporary art products in their different forms to contribute to the sense of liveability and happiness of this place. The art is fundamental for the beauty of the place. The waterfront presents artworks, above all, in the central part with the Water Mirror, which is the most used together with the retails and restaurant parts. Inserting some more sculptures or general artworks could improve the beauty, liveability and happiness of this place and then its health.

The third project intervention is to widen the game areas by inserting more playgrounds and places for games for people with disabilities and older people. The place is already equipped by sport areas, playgrounds and cycle and pedestrian paths. The sport areas could be improved with games for people with disabilities and older adults in order to allow to all to enjoy this place. These sport areas could be inserted where some are already situated, such as the Parc de Sports Saint Michel, close to the Parc des Allées de Chartres and the skatepark.

The fourth project intervention is to insert more wireless information points – such as totems, maps and so on – with the history, equipment and activities of this place. The area has a long history, and not all people know the process of regeneration through which this place has been transformed. The insertion of totems with interactive maps along the path could be useful for this aim.

The fifth project intervention is to widen the use of new technologies to give a more profound experience of the place (e.g. virtual maps, virtual experience connected to the history and local products such as the wine, etc.). New technologies are important for creating the experience of a place, relating this to a better knowledge of local products, art and architecture, contributing to the memory of it. The identity of Bordeaux is full of these elements, and to have a whole experience, the insertion of QR codes and virtual maps are useful to catch them. The physical position of the QR codes and virtual maps could be forecasted in all places along the waterfront because each stretch has its own peculiarity to be known.

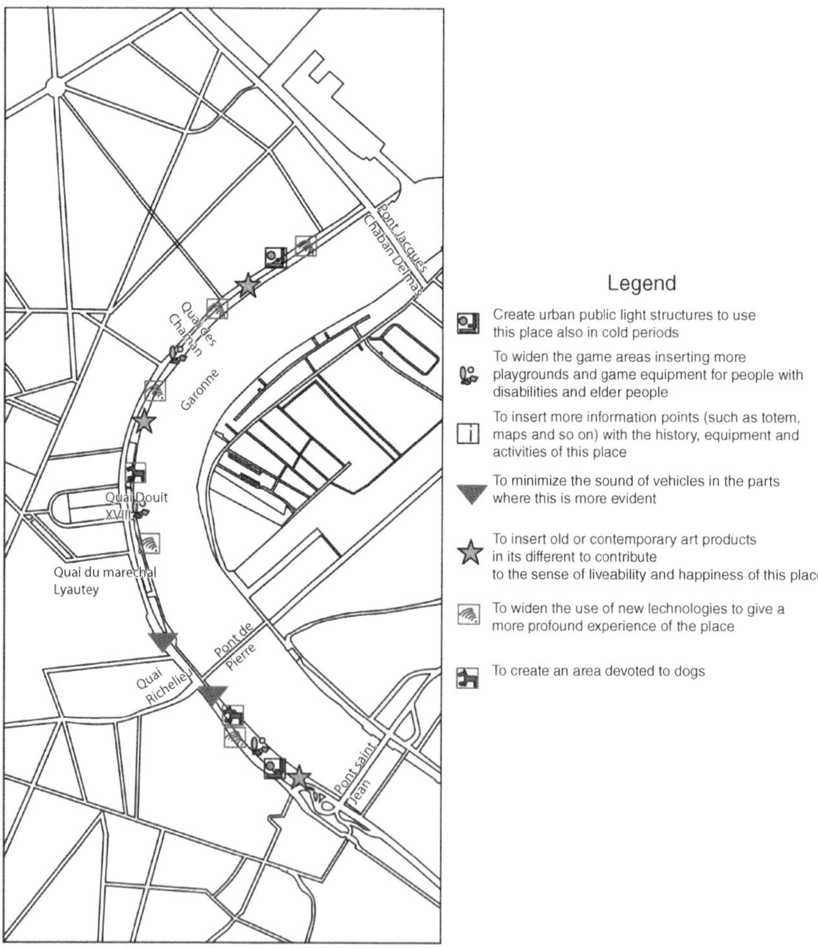


*Figure 15.11* Bordeaux riverfront, map of urban health and liveability design

The sixth project intervention is to minimize the sound of vehicles in the parts where this is more evident. In some hours of the day and in some parts of the place, the sound of vehicles can disturb the quietness of the place and its liveability. These parts include those close to the street, where light green barriers could be created (Figure 15.11).

The seventh project intervention is to create an area devoted to dogs. Dogs need their area to play with other dogs far from people who prefer to take a break or do other activities without their presence. An off-leash area and a place devoted to the enjoyment of dogs can be created close to the park and the sport areas.

## Conclusion

The beauty of this place is given by the mix of tradition and innovation that can be observed throughout the path. A point of great vital force is where the Water Mirror is located, which attracts many people and creates both feelings of joy and happiness in those who live it and in those who observe the activities carried out here, which include crossing the Mirror, playing with water and swimming (especially children). Another point of great attraction is the flower gardens that line the stretch of water. The main activities carried out along the river include walking, cycling, pausing, observing, sitting on the lawn, playing and picnicking.

The activities carried out at specific points include doing sports, booking and taking the cruise, visiting the sailing ships, buying organic and local products at the market, shopping and sitting or having something to drink or eat at the kiosks, bars and restaurants. These activities are carried out by locals of all ages and by tourists, as the area is very pleasant, healthy and liveable, with comfortable flooring and with different seating options.

The proposed activities to improve take into account some aspects that could further improve its liveability that emerged during the surveys and questionnaires.

These include creating urban public light structures to use this place also in cold periods; inserting old or contemporary art products in its different forms to contribute to the sense of liveability and happiness of this place; widening the game areas, such as inserting more playgrounds and places for games for people with disabilities and elder people; inserting more wireless information points (such as totem, maps and so on) with the history, equipment and activities of this place; widening the use of new technologies to give a more profound experience of the place (e.g. virtual maps, virtual experience connected to the history and local products such as the wine, etc.); minimizing the sound of vehicles in the parts where this is more evident; and creating an area devoted to dogs.

The regeneration process, still underway on the left bank of Bordeaux, has changed the identity of Bordeaux that was previously linked only to the idea of wine, making it more attractive, liveable and healthy. During the survey carried out to decide where to concentrate the study, outside the area in question, it was observed that the regeneration involved many parts of the historic centre, recovering many public spaces with micro interventions. These project interventions created – together with the waterfront regeneration – a network of places that starts from the historical centre and ends along the river, allowing different routes characterized by liveability and healthiness. Finally it's important that a liveable place is surrounded by other liveable sites, improving its use and promoting health behaviours as well.

# 16 Copenhagen

## The city

The case study concerns the Superkilen park in Copenhagen, the capital of Denmark, as well as the most populous, which has been involved in an extensive regeneration operation which, in the last 40 years, has transformed it from a city in decline – which has affected the port and industries since the 1980s – to one of the most liveable and happiest cities in Europe. This change was due to an innovative regeneration program that involved several neighbourhoods in a state of deterioration with the aim of creating a city with a homogeneous urban and social fabric and, above all, a resilient city: the idea is that communities have the strength to resist crises internally, adapting to them and finding a new balance. Indeed, compared to the late 1980s, there are 40% more young people and 20,500 more families with children than in the 2000s, and compared to 10 years ago, there are 37,000 more foreigners (AAVV, 2018a, 2018b).

Accordingly, international rankings and reports, including the Economist Intelligence Unit's Global Liveability Index (EIU, 2021) the Mercer Quality of Living City Ranking (Mercer, 2019) and l'ONU World Happiness Report (Helliwell et al., 2020) indicate Copenhagen as one of the happiest and most liveable cities.

## Urban planning strategies

Health is one of the factors that significantly contributes to the happiness index; in agreement, the Municipality of Copenhagen intervened both with actions relating to social assistance and with interventions aimed at encouraging healthy choices by making them the most desirable one. The idea is to implement an inclusive urban planning strategy.

With respect to social assistance, clinics to treat stress and anxiety were created. Regarding healthy behaviours, part of this policy has been aimed at prioritizing bicycles over cars, building nearly 400 kilometres of cycle paths. In addition, the traffic lights are programmed according to the speed of the bicycles, and when there is snow, the cycle paths of the driveways are cleared first, to facilitate the use of the bicycle over the car.

DOI: 10.4324/9781003098775-21

By 2025, 75% of people are projected to move by bicycle, on foot, by electric buses, or by the new subway lines under construction, and private cars will run predominantly on electricity and hydrogen. In order to improve the quality of the area, and, at the same time, that of transport, Copenhagen is also working on new environmental measures for diesel vehicles.

Another important factor to ensure liveability is that relating to housing. In order to ensure affordable prices for all, 25% of social housing is being built in all regenerated neighbourhoods so that there are social housing units in all parts of the city, avoiding gentrification.

In Copenhagen, liveability is improved by creating not only new housing, clean air, sustainable jobs and transport but also more physical spaces where people can meet, socialize and play in order to build or improve the social fabric. To create social bonds, leisure spaces such as parks have been created, as well as structures such as the Harbour Baths, where it is possible to swim in the city harbour (AAVV, 2011, 2012).

Among the strategies implemented, the main include the adaptation to climate change, the increase and enhancement of green infrastructures and public spaces, and inclusiveness.

For climate adaptation in Copenhagen, green spaces were created in the neighbourhoods capable of retaining water in cases of overabundant rainfall and then redirecting it to the city's water system. This water is then used in the summer for watering the plants. The Copenhagen Climate Adaptation Plan (AAVV, 2011), in addition to providing for the protection of Copenhagen caused by climate change, is also aimed at transforming in opportunities the risks that it can cause. The plan provides solutions to enhance the built environment and create attractive urban spaces in relation to residences and transport in order to improve the quality of life of both inhabitants and visitors of Copenhagen. The creation of climate-proof green public spaces is designed for the health and satisfaction of citizens; an important aspect in this sense is the flexible adaptation to the climate thinking about measures aimed at preventing, as well as reducing, flooding and having a diversified nature (AAVV, 2011, 2012).

The idea for a greener Copenhagen involves a city with more green and blue spaces, green roofs, trees and a city full of options for outdoor activities, where gardens, parks, rivers and lakes are connected by links and corridors.

The potential of green spaces identified by the plan is related to the realization of multifunctional green structures and measures that include absorbing rainwater to prevent flooding, creating shade and reducing energy consumption for cooling buildings, reducing air and noise pollution, promoting biodiversity, preventing stress and offering leisure opportunities.

Prevention actions are carried out, in particular, where the risk of flooding is most possible and, specifically, in the proximity of public spaces and buildings such as gardens, parks, cultural buildings, kindergartens, schools and homes for older adults.

The plan also focuses on projects that include measures that can provide the city with new inspiration and knowledge. The public spaces and existing green structures have the dual purpose of promoting climate adaptation and highlighting the identity of the places in individual spaces. The plan provides that in these spaces, in various cases included in the lists of places subject to conservation, more water is provided, and as far as possible, considering that they are cultural assets, an exception is made to the current provisions on protection.

All open spaces, such as car parks, courtyards – including school ones – vegetable gardens and various residual areas can contribute to climate adaptation through interventions that include green roofs and walls and the insertion of trees with large foliage to provide shade near schools, in common areas, in natural areas and along the roads.

In this way, continuous green networks will be created where rainwater, nature, pedestrians and cyclists share the same space. The Climate Adaptation Plan recommends ensuring areas for collecting rainwater, corridors of flora and fauna and biological diversity and recreational values.

In continuity, the new Copenhagen Municipal Plan 'World City with Responsibility' adopted in 2019 defines plans and policies for the city until 2031 (AAVV, 2019). Among these, it is foreseen that houses are located 300 metres from a green area or 500 metres from an open-air public space of 2 hectares; parking spaces are reduced by 30% except where there is a lack of public transport in order to give more space to pedestrians and cyclists; a policy is implemented to promote the use of wood in construction; 1,800 residences are built for citizens with special needs; and the good quality of housing is improved for all, including young people and students. A new, more sustainable bus network is also planned to achieve the zero-emissions target.

The Copenhagen City Plan is divided into three parts and covers the development of the city for the next 12 years. The first part of the structural policy describes the overall vision including city infrastructures, recreation and work areas. The second part concerns the guidelines, that is the administrative part that includes the cultural heritage and areas for free time, temporary use and the location of the activities. The third part concerns the rules for local planning, including housing size, density and use of undeveloped areas.

As specifically concern the public space topic, in the plan, given the forecast of an increasingly intense use of roads, cycle paths and public transport, it is planned to create a city where the car is used only if necessary, increasing the green means of transport, cycle paths and pedestrian roads, and, moreover, reducing the speed of road transport to 30 km/h.

In addition to increasing the network of cycle paths, an electric bus connection will be created to bring bicycles through the port for free and parking spaces will be increased where there is greater need. Also, by 2025 all city buses will be electric with the possibility of charging them in all depots in Copenhagen.

Another ecological form of transport is walking, which contributes to the health of the citizens of Copenhagen; the goal of the plan is to increase the number of walks in Copenhagen by 20%. To achieve this, partnerships will be established with landowners and courtyard associations to create along the route green recreational activities and areas with high-quality pedestrian connections, safer school streets and access to more stations and bus stops. Measures will be implemented to support walkability to be calculated also through the walkability index.

An adequate structure of charging points for electric cars and their free parking in the paid areas are planned to encourage their use and reduce air pollution.

To increase social equality, noise-reducing asphalt will be used in urban areas at risk, and urban spaces with good design will be created that attract people to exercise and improve mental health.

In addition, new health centres and new urban areas will be built close to the existing ones in order to reduce the distance to go to the doctor.

As mentioned earlier, there is a recreation area every 300 metres on foot from one's home. This can be made up of pocket parks, special green urban spaces, parks, lakes and the port; the type of spaces will be decided after carrying out an analysis of the needs of the inhabitants of the local community of reference.

It is then envisaged that in urban development areas the walking distance between one's home and a recreation area (over 2 hectares in size) must not be more than 500 metres.

In the local plans, to ensure the diversity of places, new areas, more connections and greater accessibility to green spaces from the city will be created through the redevelopment of urban spaces and parks and the transformation of sites and sidewalks that are mostly asphalted in green public spaces (Figure 16.1).

*Figure 16.1* Copenhagen, perspective of Superkilen area

(Source: Iwan Baan, Big)

In order to vary and improve leisure activities, the construction of not only playgrounds, skating rinks and swimming pools but also artistic installations and the transformation of existing buildings into places for new permanent and temporary cultural activities are planned. These buildings and related activities can be connected in order both to make better use of the available surfaces and to diversify and create new meeting places between various groups and neighbourhoods. The goal is to create a city for everyone, regardless of age and background.

## Superkilen

In this framework, the Superkilen project was designed. Superkilen was created as part of Prequalified Competition launched by the Municipality of Copenhagen and the Realdania association, a Danish foundation which aims to finance or support projects that improve the aesthetics and liveability of the urban environment (www.realdania.dk).The announcement required the creation of an urban park that would favour the integration of the multicultural Nørrebro neighbourhood, where 57 ethnic communities reside. The architects BIG-Bjarke Ingels Group, Topotek1 and Superflex responded by creating a place where the inhabitants themselves proposed objects from all over the world for Superkilen's furnishings, creating a liveable and inclusive place (Ciuffi, 2013).

The project was completed in July 2012 in an area of 33,000 square metres in Copenhagen that extends from Nørrebrogade to Tagensvej. The cost of the project is 7.7 million euro/11 million USD.

The design team is made up of Bjarke Ingels – partner in charge; Nanna Gyldholm Møller, Mikkel Marcker Stubgaard – project leaders; and Ondrej Tichy, Jonas Lehmann, Rune Hansen, Jan Borgstrøm, Lacin Karaoz, Jonas Barre, Nicklas Antoni Rasch, Kai-Uwe Bergmann, Gabrielle Nadeau, Jennifer Dahm Petersen, Richard Howis, Fan Zhang, Finn Nørkjær, Andreas Castberg, Armen Menendian, Jens Majdal Kaarsholm, Jan Magasanik, Henrick Poulsen, Takumi Iwasawa and Teis Draiby – the team.

Superkilen was conceived as a sort of collection of global found objects that come – as aforementioned – from 57 different home countries of the people inhabiting the adjacent area. Each object is accompanied by a small stainless plate inlaid in the ground describing what it is and where it is from in Danish and in the language(s) of its origin. The park, as affirmed by BIG, rather than perpetuating a petrified image of homogeneous Denmark, serves as a surrealist collection of global urban diversity that reflects the true nature of the local neighbourhood (Figure 16.2).

The goal is to make the Nørrebro neighbourhood become the centre of innovative urban spaces of international standard which can be an inspiration for other cities and neighbourhoods.

The park is divided into three zones that also correspond to three different colours – red, black and green – which build dynamic environments for everyday objects.

*Figure 16.2* Copenhagen, Superkilen site plan
(Source: BIG)

Mainly sports and cultural activities are located in Red Square, and it is conceived of as an extension of the activities present at Nørrebrohallen, in order to allow local residents to meet each other through games and physical activity.

In addition to sports and leisure equipment, there is a multifunctional area that can be used for markets, parades and skating rinks; the mobile stands of Nørrebrohallen can be moved there for the screenings of films or outdoor sports.

The continuity with the Nørrebrohallen and its main entrance is also achieved thanks to the integrated coloured surface in terms of colour and material; in addition, the facades are visually incorporated with the surfaces, creating a suggestive experience of a three-dimensional space for observers and users of this place.

The Red Square is bordered by a street and a building and by fences along the sides. In addition to the continuity given by the colours and materials, objects from the different cultures present in the neighbourhood are well integrated in this space.

The Black Square is the central part of Superkilen, where, on weekdays, people sit on the benches, meet around the fountain and use the standing tables to get together and play chess, backgammon and other games. The hill offers dynamism to this part of the area, as well as visual perceptions to different parts of the square. The continuity of the cycle path is guaranteed thanks to its displacement on the side of the black part in order to cross the hill (Figures 16.3–16.9).

The artistic furnishings of this part of the square include a neon sign of a dentist, bar chairs, an octopus-shaped slide, all to offer a welcoming space that gives the idea of an urban living room to all the people of the neighbourhood and their visitors.

*Figure 16.3* Copenhagen, perspective of Superkilen Red Square
(Source: Iwan Baan)

*Figure 16.4* Copenhagen, detail of Superkilen Red Square
(Source: BIG)

*Figure 16.5* Copenhagen, detail of Superkilen Red Square
(Source: Torben Eskerod)

*Figure 16.6* Copenhagen, perspective of Superkilen Black Square
(Source: Iwan Baan)

*Figure 16.7* Copenhagen, detail of Superkilen Black Square

(Source: Iwan Baan)

*Figure 16.8* Copenhagen, detail of Superkilen Green Park

(Source: Jens Lindhe)

*Figure 16.9* Copenhagen, detail of Superkilen Green Park
(Source: Mike Magnussen)

The Green Park of the area is made up of hills and green surfaces where equipment has been created for various leisure activities and sports for families and children to play hockey, badminton or basketball and to have a picnic or a break on the grass. This area is entirely green, and in addition to the landscape to give visual continuity, the cycle and pedestrian paths have been coloured in green as well.

Here there are various objects from different cultures as well, from picnic tables to barbecues, from the sports area to the dance pavilion.

To better connect this park with the entire Nørrebro neighbourhood and the other surrounding areas, the cycle paths and the entire traffic were reorganized, and new connections were created as part of a wider connection system.

Furthermore, in order to obtain a place that is liveable and well integrated into the neighbourhood, specific design strategies were implemented. The first concerns updating the tradition of the English romantic garden within the Nordic culture. This was achieved by inserting elements from foreign countries, making them an integral part of Denmark, in order to create a hybrid space between foreign and native and belonging to the local.

The second strategy was to involve the population through continuous discussion and active participation: the inhabitants of the different ethnic groups who live in the neighbourhood were asked to propose objects of their own culture so that everyone could feel an integral part of the park, contributing to its process of appropriation (Mareggi, 2020).

## Conclusion

For many years, Copenhagen has been investing a lot of energy to make the city not only more resilient to climate change but also healthier, more liveable, more inclusive and more attractive for both inhabitants and visitors.

The Plans for Climate Adaptation and Municipal Plans put a lot of attention on making the city a place that is both simple and pleasant to live and move around, creating or improving pedestrian paths, cycle paths, parking spaces, places for culture, sport and leisure in general. The quality of life is considered a fundamental factor towards which to direct policies, programs and projects.

These efforts have led Copenhagen for years to be well positioned in the rankings for quality of life.

A good practice in this direction is the Superkilen park. The Superkilen park built in the Nørrebro district of Copenhagen was born with the idea of representing the nationalities of all the people who reside there. The interesting aspect is that Superkilen represents a universal garden in which objects for street furniture, play equipment, signs and billboards are used as a pretext to reinterpret, represent and share cultures from all over the world. As Martin-Rein Cano of Topotek1 said, rather than perpetuating an idea of Denmark as a mono-ethnic people, the park reproduces an example of the cultural diversity of contemporary Copenhagen.

Furthermore, an increase in vegetation has been obtained throughout the area with the inclusion of islands with different types of trees which, due to the diversity of colours, flowering periods and the origins it represents, correspond to that of the surrounding objects in the park.

Superkilen is for all the aspects mentioned a park that has all the characteristics of liveability, healthiness and urban happiness, is frequented by people of all ages and is attractive and pleasant. Its success is determined by having carried out a place that is integrated into a wider regeneration and adaptation action that is still affecting the city of Copenhagen, in which well-being and health of citizens are considered essential in all the urban strategies and policies.

# 17 Dublin

## The area

Dublin, situated in the province of Leinster on the east coast, is the largest city and capital of Ireland. It is bathed by the River Liffey, whose banks and related bridges characterize this city and whose regeneration – together with that of the Docklands – are elements of interest and the subject of the case study.

The visits which were carried out on-site had the objective of deciding the parts to be included in the study area. Observations were carried out in the morning and afternoon hours during the week and weekend. The area that has been decided to be analysed is the Liffey waterfront between Grattan Bridge and Samuel Beckett Bridge and the area between Samuel Beckett Bridge and Grand Canal Quai-Grand Canal Bridge.

In the following, in order to better explain the results of the surveys, these are illustrated considering two parts, namely the part between the O'Connell Bridge and the Samuel Beckett Bridge and the part between the Samuel Beckett Bridge and the Grand Canal Quai-Grand Canal Bridge. The other analysis and questionnaires will be illustrated considering the area in its whole.

## People and activities

The analysis begins in the first stretch, the Liffey waterfront between Grattan Bridge and Samuel Beckett Bridge.

In the first survey of the second phase, the types of people who use this place were observed. The people who use this place were locals, residents, professionals and visitors of all ages. This place crosses the city centre and is used heavily by locals and visitors, and for this reason, many people also frequent the two sides of the waterfront.

The second phase concerns the activities carried out and relative attendance. The activities in this area include walking, observing, sitting, sitting while watching the panorama and crossing the bridges, carried out often by all people and with high frequency; running, renting a bicycle and cycling, carried out mainly by locals and residents quite often and with medium

DOI: 10.4324/9781003098775-22

frequency; and taking photos and watching the artistic sculptures, carried out quite often and with medium frequency as well.

The third survey concerned the single and mixed perceptions. The single perception concerned the visual ones of the Liffey river, bridges, trees, ancient sailing ships and the docklands; tactile ones concerned the sidewalk pavement, which is very comfortable, and the touch of the realistic and "engaging" sculptures situated close to the benches; the smell ones concerned the river and, in the spring period, the flowering trees; and the acoustic ones concerned, above all, the means of transport which is louder in the tighter stretches and in the road junctions. The mixed perceptions related to this stretch include the serenity, comfort and liveability in the whole stretch and of chaos in the parts with road junctions.

With respect on the elements which contribute to the health, liveability and happiness perception, these include the wooden boardwalk, benches and seats which face the river; trees and little street gardens and flower pots; bike racks and bikes for rent; signs with the history of the place; light poles and kiosks with outdoor tables and chairs; and a comfortable pavement and artistic bird tracks and footprints embedded in the pavements near the O'Connell Bridge (Figures 17.1–17.9).

The second stretch concerns the area between the Samuel Beckett Bridge and the Grand Canal Quai-Grand Canal Bridge.

*Figure 17.1* Dublin, O'Connell Street with the artistic bird tracks and footprints

*Figure 17.2* Dublin, Liffey Riverfront perspective

*Figure 17.3* Dublin, Liffey Riverfront detail with the *Linesman* sculpture

*Figure 17.4* Dublin, Liffey Riverfront detail with the *Jeanie Johnston* historical sailing ship

*Figure 17.5* Dublin, Liffey Riverfront, a café

*Figure 17.6* Dublin, Liffey Riverfront, Samuel Beckett Bridge

*Figure 17.7* Dublin, Grand Canal Square perspective

*Figure 17.8* Dublin, Grand Canal Square perspective

*Figure 17.9* Dublin, Grand Canal Square detail

The people who use this stretch are the same of the previous stretch with a major presence of visitors in the public space in front of the Bord Gáis Energy Theatre and around it.

With respect to the activities carried out, these include walking, crossing the bridges, sitting and sitting on the benches to watch the panorama or take a break, which were carried out often by all people and with a high frequency; running, renting a bicycle and cycling, canoeing and kayaking in the Liffey, which were carried out mainly by locals – except cycling, which is mainly carried out by visitors – quite often and with medium frequency; and taking photos and watching the Bord Gáis Energy Theatre and the public space around it, and watching the panorama in general are carried out by all quite often and with a medium degree of frequency as well.

Regarding the perception surveys, the single perceptions include the visual ones, which concern the Grand Canal, the Docklands, the Bord Gáis Energy Theatre and the opposite side of the river; the sound ones, which concern the voices of people, the movement of ships in the canal and the mean of transports; the tactile ones, which concern the benches, the green carpet and the lawns; and the smell ones, which concern the river and the green spaces.

The mixed perceptions include joy, surprise and happiness – in Grand Canal Square – serenity, healthiness and liveability in the other parts.

With respect on the elements which contribute to the health, liveability and happiness perception, these include benches and seats which face the river, trees, bike racks, bike rental, canoe and kayak rental, signs with the history of the place, light poles in the waterfront and the comfortable pavement; the particular benches, the red light sticks, the red-and-green carpet and the comfortable pavement in Grand Canal Square.

After the surveys, 120 people – residents aged between 20 and 65 years – and visitors from all over the world, mainly from Europe and the US – aged between 30 and 70 years were administered a questionnaire. The language used was English. People generally answered the questions posed with pleasure.

The following questions were asked:

1. Does this place give you a feeling of happiness or sadness/liveability or discomfort/healthiness or unhealthiness?
2. What are the elements that give you the above sensations?
3. What are the main facilities that give quality to this place?
4. What are the activities that you do in this place, and how often?
5. What do you think about the presence of many or few people here? Do you think that it is capable to improve the pleasantness or unpleasantness of the place?
6. According to the place's health, liveability and happiness, what could be done in order to improve this place?
7. What is a healthy/happy/liveable place that you remember in this city or elsewhere?
8. How might the weather condition influence the perception of the health, liveability or happiness of this place?

To the question, "Does this place give you a feeling of happiness or sadness/liveability or discomfort/healthiness or unhealthiness?" 100% of locals and visitors independent of age answered liveability, with particular attention to the parts of these areas in which there is more space to sit, walk and/or watch the panorama.

To the second question, "What are the elements that give you the above sensations?" 80% of locals of all ages answered the walk on the waterfront and the bridges and the sculptures; the other 20% of locals of all ages answered the Samuel Beckett Bridge and Grand Canal Square. Of visitors aged between 30 and 65 years, 70% answered the walk on the waterfront, the Samuel Beckett Bridge, the sculptures, the flowerpots and the design and furniture of Grand Canal square, while those between 20 and 30 years mainly answered the Samuel Beckett Bridge and Grand Canal square along with the Bord Gáis Energy Theatre; the other 30% of visitors of all ages answered the panorama in front of Grand Canal Square and the area as a whole.

To question 3 – "What are the main facilities that give quality to this place?" – 100% of locals and visitors of all ages gave more than an answer. Eighty per cent of them answered the benches, the sitting areas, the bikes to rent, the bridges, the kiosks with open-air tables and chairs, the urban furniture of Grand Canal Square and the Bord Gáis Energy Theatre; the remaining 20% added to these elements the light poles, the flowerpots and the artistic bird tracks and footprints embedded in the pavements of O'Connell Street.

To the question "What are the activities that you act in this place, and how often?" 80% of locals and visitors of all ages answered walking, sitting, sitting at the cafes and watching the panorama; of these the 40% of people aged between 30 and 50 added cycling and running. The remaining 20% of locals and visitors of all ages answered watching and taking a break.

To question 5 – "What do you think about the presence of many or few people here? Do you think that it is capable to improve the pleasantness or unpleasantness of the place?" – 80% of locals and visitors answered that the presence of too many people in the strictest parts of the waterfront could be unpleasant, while the presence of many people on Grand Canal Square is agreeable because the area is wider. The other 20% answered that in general, the presence of many people improves the pleasantness of the place. Of both locals and visitors aged between 20 and 30 years, 100% answered that the presence of a few or many people do not change the perception of the place.

To the question, "What could be done in order to improve this place according to the place's health, liveability and happiness?" 80% of locals of all ages answered reducing vehicle traffic close to the road junctions and adding an off-leash area for dogs; the other 20% answered adding some equipment for gymnastic exercises and more trees in the final part of the waterfront. One hundred per cent of visitors answered more actions together, namely reducing vehicle traffic close to the road junctions, improving information about the area and adding more green areas.

To the question, "What is a healthy/happy/liveable place that you remember in this city or elsewhere?" 70% of local people of all ages mainly answered that it is a unique place; the remaining 30% answered, beyond the areas in the analysis, Temple Bar, Trinity College or Saint Patrick's, always in Dublin.

Of visitors of all ages, 100% answered the Liffey waterfront and the Grand Canal area in Dublin, adding that in general, this is a liveable city. The 20% of those interviewed added the New Tate area in London, Central Park in New York, Parc Rives de Seine in Paris and Ostermalm in Stockholm.

Finally, to the question, "How the weather condition might influence the perception of liveability or happiness of this place?" 70% of locals and visitors answered that the weather condition may influence the perception of the place in Grand Canal Square, while perception of the Liffey waterfront does not change in a particular way. The remaining 30% answered that the cold weather could change the liveability and health of both the waterfront and Grand Canal Square.

## The transformation of the area

The next step of the analysis consists of traditional analysis. The regeneration process of Dublin began in the late 1980s with a transformation that provided Dublin with new urban areas that were added to the creation – between the 18th and 19th centuries – of a system of services and public spaces, residential neighbourhoods, the port and the Grand Canal (Figure 17.10).

*Figure 17.10* Dublin, Temple Bar perspective

The works relating to the 1980s are in particular related to large infrastructure, including the Dublin Area Rapid Transit metropolitan railway. Since the mid-1990s, physical and economic growth has increased dramatically, with the construction and transformation of neighbourhoods, including the historic Temple Bar, the Docklands, business centres, residential areas, the first tramway, O'Connell Street and several public spaces (Moore-Cherry and Vinci, 2012; Dublin City Council, 2003; Dublin Docklands Development Authority, 2000, 2006; Gospodini, 2002; Lawton et al., 2010; Montgomery, 1995).

The Temple Bar – the area behind the Liffey waterfront – regeneration project was born thanks to the will of the local population, who fought against the demolition of this historic district which, for years, had been affected by severe urban decay that also compromised the surrounding areas.

To avoid demolition, many buildings were taken over and rented at low cost to artists, architects and small shops, convincing the public to change their minds about that area. The transformation of the district has led to the creation of a district of great cultural interest, as well as a place for meeting and socializing for locals and tourists, with the presence of numerous cultural institutions, resulting in a renewal of the surrounding areas as well.

Regarding monuments, the main ones include the Custom House, the Famine Memorial statues, the *Linesman* sculpture, the Four Courts and the reproduction of the *Jeanie Johnston* historical sailing ship. There are many bridges, including Ha'penny Bridge, O'Connell Bridge, Grattan Bridge and the Millennium Bridge.

In 1997, the Docklands regeneration project began and has among its most interesting elements the redesign of the pedestrian street along the River Liffey, the Samuel Beckett Bridge built by Calatrava, the Bord Gáis Energy Theatre built by Libeskind and the square designed by Martha Scwartz. In the same year, a 650-metre pedestrian route was realized on the north side of the river to create a footpath with benches and kiosks separated from the traffic chaos.

## The liveability and healthiness factors

The last part of the analysis concerns the identification of place and activities which make the place liveable, healthy and happy from the user's point of view.

The area that was analysed is an area quite recently regenerated. The waterfront of the River Liffey is a liveable and healthy place that offers people the opportunity to do different types of activities. The river serves as a scene for walking, cycling, running, stopping for a break, having a hot drink at the café, gazing at the view, sitting, relaxing and taking a sightseeing tour by boat. Here there are seats, benches of wood and stone of different shapes and oriented in different ways, which allow a view of the river and for conversation, trees, green flower beds, decorative flower pots and bike-parking racks. The stretch with the wooden walkway separates the walk from the

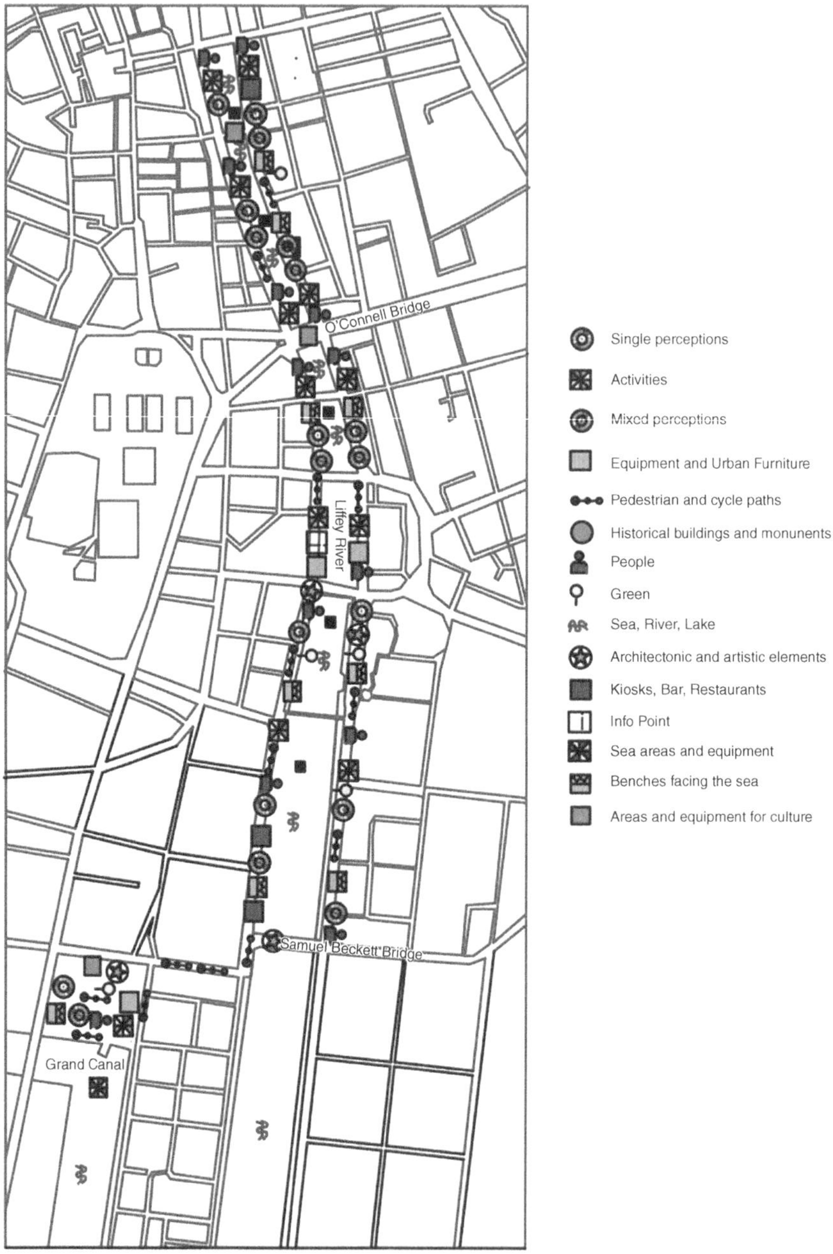


*Figure 17.11* Dublin, Liffey Riverfront, map of urban health and liveability analysis

noise of road vehicles, giving people an even closer view of the river with the possibility of sitting on the benches. The bronze statues along the waterfront, as well as artistic bird tracks and footprints embedded in the pavements at O'Connell Street, the visual perceptions of historic buildings such as the Custom House, the bridges and the sailing ships contribute not only to the perception of the liveability of the place but also to the enhancement of its identity; also, the outdoor exhibition panels that illustrate the history of the area constitute an important factor for the liveability of this place.

The long river ends at the Docklands, where the second stretch of the study area is located, between the Samuel Beckett Bridge and Grand Canal Quai-Grand Canal Bridge. This part is characterized by a pedestrian stretch in continuity with the long river but is wider, with seats, trees and bike-parking racks. Here people walk, take a break, observe the river, run, and ride their bikes.

The walk reaches the Grand Canal Plaza with contemporary architecture and public spaces that offer a perception of surprise and almost discovery, heralded by the Samuel Beckett Bridge. The Grand Canal Plaza is made up of a contemporary design with seats, paths marked by pavement and stylized red lampposts. Here people walk, take a break, ride and rent bicycles and take pictures of the square, theatre and panorama. The perception is of liveability, healthiness and happiness (Figure 17.11).

The whole area is routinely frequented by locals and tourists of all ages, both for its beauty and liveability and for its centrality and proximity to areas of interest, such as the historic Temple Bar district and the Docklands.

## Designing health and liveability

The check of the degree of healthy, liveability and happiness is carried out through the 25 principles of urban health and liveable design concerning the Charter. By the check of the principles, those which could be improved include (7) to create suitable spaces for dogs and domestic animals; (9) to minimize or eliminate the noise generated by public transport; (10) to improve suitable cycle lanes; (15) to have the possibility of using the space in different weather conditions and seasons, contributing at the same time to its good state of maintenance; (17) to both allow and promote different types of functions such as games, breaks and walking; (18) to facilitate gymnastic activities with the presence of small equipment or a designated space; (20) to encourage the presence of art in its different forms; (22) to promote participation, namely the feeling of being able to contribute to the life of that place increasing the sense of belonging; (24) to promote the educational function which a place has – for example clearly displayed information about history of the place and so on or suitable ways to use it – increasing its intrinsic value; and (25) to facilitate the use of new technology to increase the knowledge of its intangible values and history, offering a more profound experience of the place.

The items of the questionnaire derive from the principles which could be better satisfied. About 120 locals and visitors from all over the world – mainly from Europe and the US – between 20 and 65 years of age were administered the questionnaire. The language used was English and people were willing to give answers to the questions.

In the following, questions and answers are reported.

1. What about adding a cycling- and a pedestrian-dedicated lane to improve the connection between the walk of the Liffey River and Grand Canal Square?
2. What do you think about inserting playgrounds?
3. What do you think about inserting artworks such as sculptures to improve the presence of art in this place?
4. What about adding an area for dogs?
5. What about adding information about the area, as well as using new technologies?
6. What do you think about adding some equipment for gymnastic exercises?
7. What about improving the presence of green spaces?
8. What about identifying suitable sites for participation events?
9. What do you think about creating urban public light structures to use this place also in very cold or warm periods?

With respect to the first question – "What about adding a cycling- and pedestrian-dedicated line to improve the connection between the walk of Liffey River and Grand Canal Square?" – 80% of people – both locals and visitors of all ages – answered that they agreed, adding the importance of having different paths for cyclists and pedestrians. The other 20% answered that they agreed and that the differentiated paths should all be realized in the waterfront area.

Regarding the second question – "What do you think about inserting playgrounds?" – 70% of local interviewees of all ages answered that they agreed with this idea. The remaining 20% of locals answered that they agreed, adding that the more suitable part could be the larger part of the wooden walkway on the Liffey River. All visitors of all ages answered positively.

With respect to the third question – "What do you think about inserting artworks such as sculptures to improve the presence of art in this place?" – 80% of both locals and visitors answered that they agreed with this idea. Of these, 40% of younger adults added that art could include street art such as murals. The remaining 30% of interviewed – independent of the age – answered that they agreed, adding that the artwork should be in line with the identity of the place.

Regarding the fourth question – "What about adding an area for dogs?" – 80% of people – both locals and tourists of all ages – answered that they would like to have an off-leash site in this area for dogs. The remaining 20% – mainly locals aged between 30 and 60 years – added that the off-leash site should be located close to Grand Canal Square because along the riverfront, at moment, there is not sufficient space for this.

As regards the fifth question – What about adding information about the area, also using new technologies? – 100% of interviewed people answered that they agreed with inserting more information about the history of this place and, in general, other kinds of information about the activities which could be carried out there. Of these, 50% of people added that this information could be inserted in QR codes or apps (Figure 17.12).

With respect to the sixth question, "What do you think about adding some equipment for gymnastic exercises?" 70% of interviewees – locals and visitors of all ages – answered positively, although locals had different ideas about the typology of equipment. The other 30% of people answered that they agreed with this idea, adding that Grand Canal Square and the Docklands area could be the most suitable place for the equipment.

As regards the seventh question, "What about improving the presence of green areas?" 90% of locals and visitors of all age answered that they agreed; locals added that while in some parts of the area, there are many trees and green spaces, in other parts, there are not and that they should be introduced. The remaining 10% of people answered that they would like to see more flowers than green in general.

Regarding the eighth question – "What about identifying suitable sites for participation events?" – the 70% of locals of all age answered positively adding that these could be organized in the walkway or in Grand Canal Square. The remaining 30% of locals answered that it is not necessary to identify specific sites, all of the area could be used for participation events. All the people interviewed generally agreed with the organization of participative events, although they did not know which sites would be most suitable.

With respect to the ninth question – "What do you think about creating urban public light structures to use this place also in very cold or warm periods?" – the 90% of interviewed people – locals and visitors of all ages – answered that they agreed with this idea. The remaining 10% agreed, adding that the light structures should be realized in wood or natural materials.

Checking social networks and websites with suggestions, pages on Instagram include grandcanaldock, with 162 posts and 300 followers, and bordgaisenergytheatre, with 1,093 posts and 20,300 followers; the hashtags #dublindocklands, with 10,659 posts; #grandcanaldublin, with 2,783 posts; and #liffeywalk, with 40 posts. With respect on Facebook, there is the Bord Gáis Energy Theatre official page, with 93,677 followers, and on Twitter, there is @dublindockland1 page.

With respect to TripAdvisor, River Liffey has 765 reviews; Liffey Boardwalk, 71 reviews; Samuel Beckett Bridge, 643 reviews; O'Connell Bridge, 418 reviews; Ha'penny Bridge, 2,370 reviews; and Bord Gáis Energy Theatre, 865 reviews. The reviews are very positive on the beauty of this place, only reporting that in some hours the boardwalk is full of people and the Liffey waterfront and its bridges are full of cars and people. In Booking.com website, the hotel reviews often mention the Liffey River and the Docklands positively.

From the information gathered in the previous phases of the project, the following project interventions were identified.

The first project intervention is to create a cycling- and a pedestrian-dedicated lane. The Liffey waterfront is a place used by many people, and creating both a cycle and a pedestrian line could make this place still more liveable. Where the street dimensions are not suitable, this project intervention could be carried out with the enlargement of the current pavements and consequent reduction of the car lane. The dedicated bike and pedestrian lanes could continue until Grand Canal Square and along the Docklands.

The second project intervention is to create new spaces, widen the game areas and insert playgrounds and places for games for people of all ages and those with disabilities. Also in this case, these could be inserted in the wider parts of the area, including the Liffey walkway and Grand Canal Square. Playgrounds could also be of small dimensions and made by natural materials. This project intervention could further increase the liveability and healthiness of this place.

The third project intervention is to insert artworks to improve the presence of art in this place. Art products could be of different kinds, including sculptures, video and murals that could be inserted in the whole site and could be permanent or temporary. These, together with those which are already present in the area, could be engaged to further improve the liveability of the place.

The fourth project intervention is to add equipment for gymnastic exercises. This could be inserted in the part of the waterfront that is wider, including the Liffey walkway and in Grand Canal Square. The equipment could allow people of different ages and abilities to undertake physical activities and further improve the perception of the healthiness of this place.

The fifth project intervention is to add an area for dogs. This means to insert an off-leash space in which dogs can run and play freely in a dedicated area of the waterfront and could be carried out with an enlargement of the pavements or the pedestrianization of the whole or parts of the Liffey River. Alternatively, an off-leash area could be inserted in Grand Canal Square.

The sixth project intervention is to minimize the sound of vehicles. Along the Liffey riverfront the sound of the vehicles in some hours of the day is quite loud; a speed limit reduction for means of transport, the pedestrianization of some parts or the whole waterfront and the introduction of green barriers are the project interventions which could minimize the impact of the vehicular traffic.

The seventh project intervention is to insert public light structures to use the place in all the seasons. The structures should be mobile, made with natural material and be in continuity with the identity of the place. Furthermore, these could be used in both very warm and cold periods, further improving the use of the place all around the year.

The eighth project intervention is to add information about the area, and it is strictly related to the idea of inserting wireless information points. Although some information points are already present in the area, adding some more information with the history, equipment and activities of this place could improve its liveability and happiness. This information could be inserted using totems, QR codes, a dedicated app and so on.

The ninth project intervention is to identify suitable sites for participation. This is important to create major sense of belonging of the place and could be temporary, namely according with a general identification of the suitable sites, could change in relation to the kind of activities. These could include the wider part of the riverfront, the boardwalk and Grand Canal Square.

The tenth project intervention is to insert or enhance the green. The Liffey riverfront and Grand Canal Square present trees and green areas in different forms. In some parts where the pavement is of small dimensions, there are no trees or green spaces; the insertion of trees, green spaces or flowerpots could further improve the health and happiness of this place.

*Figure 17.12* Dublin, Liffey Riverfront, map of urban health and liveability design

## Conclusion

The Dublin areas analysed are places of great interest because there are elements of tradition and elements of innovation. The chosen route starts from the Grattan Bridge and ends in Grand Canal Bridge. In this path, the River Liffey constitutes the natural element that characterizes the place and determines its activities and the flow of people and means of transport on the two banks, on the bridges and along the docklands.

The Docklands regeneration project led to the redesign of the pedestrian street along the River Liffey, the Samuel Beckett Bridge, the Bord Gáis Energy Theater and the square designed by famous architects and landscape architects. These projects – together with the previous ones which include Temple Bar, Docklands and O'Connell Street – renovated the urban image of Dublin, improving its attractiveness.

Furthermore, the new wooden boardwalk, a light wooden structure that runs along the riverfront, allows those who come here to experience a walk characterized by the beautiful cityscape view. After crossing the new Samuel Beckett Bridge, people reach – through the Grand Canal Quai – Grand Canal Square, which is of contemporary design as well. Here the design of the public spaces, the Bord Gáis Energy Theater, the open view of the River Liffey and the docklands offer a great perception of liveability, healthiness, and happiness.

The good quality of the materials used for the streets, the good maintenance, the presence of trees and greenery and the sculptures contribute to this general perception.

The entire route is experienced by locals of all ages and by visitors from all over the world for walking, cycling, running, stopping for a break, having a hot drink at the café, gazing at the view, sitting, relaxing and taking a sightseeing tour by boat.

Some aspects can be improved to make this place even more liveable and inclusive, which include creating a cycling- and a pedestrian-dedicated lane, widening the game areas, inserting artwork to improve the presence of art in this place, adding equipment for gymnastic exercises, adding an area for dogs, minimizing the sound of vehicles, inserting public light structures, adding information about the place, identifying sites for participation events and inserting or enhancing the green spaces.

These project interventions have as a *file rouge* to give more space to people and less to cars, improving the health of the site while maintaining its place identity.

# 18 Hamburg

## The area

The case study concerns the public spaces of HafenCity in Hamburg, northern Germany, a 155-hectare area, involved in a wide operation of urban regeneration which started in 1998. Surveys, observations and questionnaires were carried out over a year and in different periods, both during the week and the weekend and during daytime and afternoon hours.

The first phase consists in the choice and definition of the study area. Three site inspections were carried out on the whole study area and in the surroundings to identify the sites of interest for the experiment aims. The places which were chosen for the case study concern the HafenCity main public spaces which, in their totality, give continuity to the area, namely the boundaries which were identified are Hubenerstrasse on the southern part and Am Sandtorkai on the northern part, including Grasbrookhafen and Sandtorhafen on the western part and Am Sandtopark on the eastern part.

The second phase is characterized by the observation of the characteristics of the place. Surveys on the activities and relative modalities, on the perceptions and elements which contribute in different ways to the health and liveability of the place were carried out.

Regarding the first survey, the typology of people (locals, visitors, professionals, etc.) and activities (passing by, working, taking breaks, etc.) were observed. These activities are measured from the quantitative point of view, namely measuring how the specific activity is present in that place and influences its liveability and use.

In order to illustrate the activities, the place is divided into seven stretches: (1) Grasbrook, (2) Marco Polo Terraces, (3) Dalmannkai Promenade, (4) Vasco de Gama Platz, (5) Magellan Terraces, (6) Überseeboulevard and (7) Lohsepark.

## People and activities

Grasbrook playground, Marco Polo Terraces, Vasco de Gama Platz and Magellan Terraces public spaces are mainly connected through the Dalmannkai Promenade and Überseeboulevard, while Lohsepark is connected to Brooktorhafen in the north by the Ericus Bridge.

DOI: 10.4324/9781003098775-23

Surveys, which started in Grasbrook and ended in Lohsepark, were carried out in the morning and afternoon hours both on weekdays and on the weekends. An average of the results obtained in the different surveys is illustrated; if particular differences during a specific day or hour were observed, this information is reported.

Regarding activities which are present in the Grasbrook playground, the main ones concern playing, picnicking, taking a break, educational, socialization and passer-by activity. The playing activity is carried out in a frequent manner by children of different ages, including nursery or primary classes – in the morning hours – or with adults in different hours of morning and afternoon, in both cases frequently.

For the children in primary school, the leisure activity is often connected to the educational ones devoted to illustrating the new HafenCity area and its meaning in relation to the reconstruction of the maritime identity through the games present in the Grasbrook playground.

The next area is constituted by the Marco Polo Terraces, where the activities present included observation, break, socialization, lunch break, passing by and cycling. The activity which is mainly present is the break carried out by people of ages between about 20 and 50 years both during the morning and afternoon hours in a frequent manner. In addition to this activity, lunch or coffee breaks are equally present, and socialization happens in groups of two and three people. The activities of passing by and cycling are also present and are carried out in all the HafenCity public spaces.

Public spaces continue with the Dalmannkai Promenade, where it is also situated along the Vasco de Gama Platz. In the Dalmannkai Promenade, the most frequent activities include passing by, cycling and taking a break in one of the different places for sitting, from where it is possible to admire the sea from many points of view. The average age of people who carried out these activities is between 20 and 60 years and the typology is mainly constituted by locals, residents – mainly workers of the area – and tourists of many countries. Furthermore, there is the presence of some bars, shops with souvenir shops, and restaurants which increase the activity of breaking, although if in private places. On the Vasco de Gama Platz, the most present activity is that of taking a break, carried out in a frequent manner on the part of the square overlooking the sea and where there are many deck chairs.

Continuing the observation of Dalmankai Promenade, there are the Magellan Terraces, where the activities which are mainly present include taking a break, cycling, walking and observing. These activities are carried out mainly by people aged between 20 and 50 years.

Taking a break takes place both on the benches and the terraced steps. From there, the break is improved by the observation which takes place in different parts of the surrounding space. Furthermore, the urban landscape is enriched by the nearby museum of music on one side and the old sailing ships– visitable – on the other and the different buildings which create a very diversified skyline.

In Überseeboulevard, the prevalent activity is walking, followed by taking a break in the different bars and restaurants of the area. Both activities are very frequent and are carried out by people of all ages and include locals, residents and tourists. The pace with which these activities are carried out is generally moderate and slow, offering a very agreeable perception of that place.

Finally, surveys were carried out in Lohsepark, the wider green area of HafenCity. Here the main activities are walking, running, cycling, playing, taking a break and sitting on the steps. These activities are carried out in a frequent manner mainly by locals and residents, while to walking, taking a break and sitting on the steps are carried out also by visitors. The pace is slow, contributing to the pleasant perception of the site. Benches and mobile chairs of different kinds, game areas and terraces with steps constitute the main equipment for this green park.

Then the second survey concerning the single and mixed perceptions was carried out. Regarding the single perceptions, these are all involved but taste – in particular, in the Grasbrook playground.

The main perceptions – all agreeable – include the sound ones given by the voices of people who live in this place and by the water and wood of the games; the tactile, given by different materials – sand, water, meadow and wood – which are used for the ground of the park; the visual, given by the scenic games with which the park is built; and the smell once, given by the smell of the natural material used for the features of the park and by the sea situated not so far away.

In the Marco Polo Terraces, Dalmannkai Promenade, Vasco de Gama Platz and Magellan Terraces, the single perceptions which are felt regard the visual ones concerning the sea and surrounding constructions, including the views of the Elbphilharmonie and the old sailing ships on the Magellan Terraces; the acoustic ones, concerning the sounds of voices and footsteps of people, low and agreeable, and, in some moments, the sound of the passing ships; the tactile ones, concerning the different pavements of the walks and sittings – wood, asphalt, concrete and meadow; and the smell once, concerning the sea, and – in particular, close to the Dalmannkai Promenade, the smell of coffee and food.

Regarding observations of the Überseeboulevard, the main perceptions include the acoustic ones, due to the sounds of means of transport and building sites – the area is still in development – and the taste ones, due to the food of the many restaurants and cafes in the area.

Regarding the mixed perceptions, the most frequent are those of cheerfulness, joy and serenity. The feeling of cheerfulness is perceived mainly in the Gransbrook Park due to the children's games built with natural material (Figures 18.1–18.9).

The feeling of joy and serenity can be experienced in all the public spaces, as testified by both the low or moderate pace and the high frequency with which these activities are carried out.

*Figure 18.1* Hamburg, HafenCity area reached from a bridge

*Figure 18.2* Hamburg, HafenCity, perspective of the Speicherstadt warehouse

*Figure 18.3* Hamburg, HafenCity, Magellano Terraces

*Figure 18.4* Hamburg, HafenCity, stairs to observe the panorama

*Figure 18.5* Hamburg, HafenCity, Vasco de Gama Platz people on deck chairs

*Figure 18.6* Hamburg, HafenCity, Vasco de Gama Platz, the path

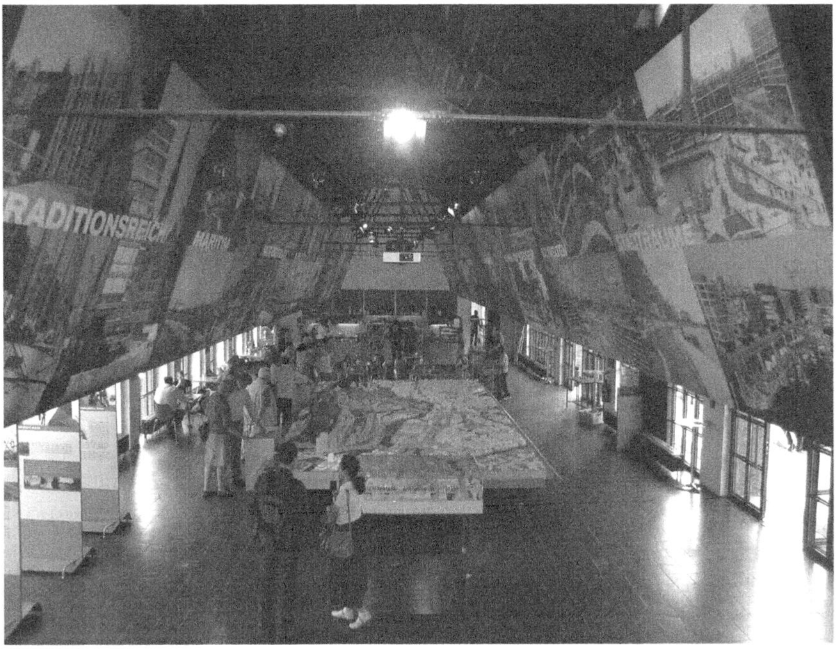

*Figure 18.7* Hamburg, HafenCity, the HafenCity InfoCenter

*Figure 18.8* Hamburg, HafenCity, view of De Elbphilarmonie from the Magellano Terraces

*Figure 18.9* Hamburg, HafenCity, the Grassbrok garden

Finally, with respect to Lohsepark, the single perceptions – all pleasant – concern: the visual ones which include the Elbe, Ericus bridge, Automuseum Prototyp building, and the rows of trees; the acoustic ones, which include the voice of people; the tactile ones, concerning the lawn grass, the chairs, the benches and the steps; and the smell ones, which regard the trees and the lawn grass.

With respect to the mixed perceptions, the most frequent are those of healthiness, liveability, joy and serenity and the pace with which the activities of the park – relaxing, making physical activities, playing, observing – are carried out is generally low, and the frequency is medium-high.

The third survey was that of elements which contribute to the perception of health, liveability and happiness.

The elements of the area which contribute to these perceptions include the presence of large and of quality design public sittings in different points: from those built using concrete to the deckchairs to the many stairs, creating and favouring multiple choices of open-air activities which vary from the observation of the place to the lie down for sunbathing – during the warm months – and to the eating the lunch or having a coffee break.

The sittings are located in well organized and of agreeable design paths posed on different levels which allow both walking and cycling. The presence on the Magellan Terraces of the historical sailing ships contribute to creating a welcoming atmosphere, recalling the ancient maritime memory of the place. The historical memory is also created by the industrial warehouses which were recovered, contributing to the beauty of the place. Old and contemporary – the latter given by the new architectures and the Herzog-De

Meuron Elbphilarmonie – create an image of continuity with the whole city of Hamburg and then of cultural sustainability as well.

Phase 3 concerned the questionnaire on-site to the users of the place devoted at comprehending the elements which make this place liveable and healthy.

The questionnaire was administered to about 100 people in English, both local and tourists; questions were carried out mainly to people who were taking a break on the benches or on the lawn in the different stretches of the HafenCity. All the people interviewed answered to the questions, even though with different times dedicated to the answers.

The following questions, and their answers, are reported.

1. Does this place give you a feeling of happiness or sadness/liveability or discomfort/healthiness or unhealthiness?
2. What are the elements that give you the above sensations?
3. What are the main facilities that give quality to this place?
4. What are the activities that you do in this place, and how often?
5. What do you think about the presence of many or few people here? Do you think that it is capable to improve the pleasantness or unpleasantness of the place?
6. According to the place's healthiness, liveability and happiness, what could be done in order to improve this place?
7. What is a healthy/happy/liveable place that you remember in this city or elsewhere?
8. How might the weather conditions influence the perception of the healthiness, liveability or happiness of this place?

To the question, "This place gives you a feeling of happiness or sadness/liveability or discomfort/healthiness or unhealthiness?" 100% of those interviewed, both locals and tourists, answered liveability, even though with a different meaning depending on the different stretches of the area. The terms *happiness* and *liveability* were used mainly for Grasbrook park, while for the other spaces, the terms which were used include *joy*, *serenity*, and *health*, recalling agreeable sensations perceived in HafenCity.

Regarding the second question, "What are the elements that give you the above sensations?" answers regarded different elements. For the locals, 100% answered the possibility of enjoying the sea, the public spaces, bars and restaurants and then the easy connection with the centre of Hamburg. Locals who were interviewed were both people who work in the area (70%) and people who work and live elsewhere (30%). Regarding tourists, about 80% answered that the elements concern the new architecture, the old sailing ships, the boats which pass from there, the bars and the presence of many places for sitting. The remaining 20% answered in a more generic manner, affirming that this area is well maintained and organized.

To the question, "What are the main facilities that give quality to this place?" 80% of locals answered the playground, the deck chairs and the other sittings and the presence of the university with many outdoor public spaces. Of tourists, 80% added to these the little modern sculptures, the sailing ships and the HafenCity InfoCenter with information concerning the different phases of urban transformation and the model of the whole area. The remaining 20% of interviewees, both local and tourists, answered the public and private ships for the tour of HafenCity by the sea.

To the question, "What are the activities that you act in this place and how often?" answers were different but consistent with the many possibilities of activities which the place offers. Of locals who live there, 80% answered that they go to HafenCity to work and use the public spaces during their lunch break. The remaining 20% of locals answered that they go to HafenCity to walk, go to the playground – in particular adults with children and students, go to the bar and take a break, with a weekly frequency or more often thanks to the means of public transport which easily connects the area with the city centre. One hundred per cent of tourists answered that HafenCity is a new and very beautiful area of Hamburg which they visit in one or two days, taking a break, strolling in the public spaces and eating or drinking in the cafeterias or restaurants.

To the question "What do you think about the presence of many or few people here? Do you think that it is capable to improve the pleasantness or unpleasantness of the place?" 100% of both locals and tourists answered that the area is wide and the presence of a medium quantity of people increases the pleasantness of the place, giving an atmosphere of vitality to the area. The different functions present in HafenCity offer a variety of uses in large spaces by people of different ages, proveniences and professions.

Regarding the question "According to the place healthy, liveability and happiness what could be done in order to improve this place?" the answers given by about the 100% of locals were nothing, the place is liveable as it is. Some of them (10%) added the quicker completion of the work sites which can obstacle the easy use of the place, while another little part (8%) added that more green spaces could be inserted in some parts. Tourists have mainly answered (80%) that they would not change anything as well. The remaining 20% proposed some improvements, including the organization of more artistic events, the insertion of more equipment or activities of interest for older adults and the insertion of areas for repairing on rainy and cold days.

To the question "What is a healthy/happy/liveable place that you remember in this city or elsewhere?" answers concerned many places given the variety of people who were interviewed.

Sixty per cent of locals answered the Rathaus Platz of Hamburg, which represents a sort of symbolic place of the city; the remaining 40% answered the square – also in Hamburg – with the swans and ducks to be admired. Tourists given the more diverse answers, referring to their city of belonging and to the connection of those with the sea or the river, including the Boston waterfront, the London docklands and the Parc Rives de Seine in Paris.

Finally, to the question, "How the weather conditions might influence the perception of healthy, liveability or happiness of this place?" all locals and tourists answered that any area is used the whole year, because there are visitors, professionals, students and residents. In periods of cold and rain, the perception of the place, as in many cities, changes for many people due to the difficulty of staying outdoors for a long time. In that period, the spaces which are used more are mainly the private ones.

## The transformation of the area

The traditional analysis (phase 4) that emerged from the area's history played a fundamental role in the waterfront regeneration. Although only a partial reconversion of the harbour warehouses was carried out – responsible for the warehouse district: HHLA Hamburger Hafen und Logistik AG – the project has restored the maritime vocation of the historical city and is strictly connected with the city.

Hamburg is Germany's main port and one of the major European ports. Navigable rivers and canals and a dense road and railway network connect it to many locations in Central and Northern Europe. Its origins go all the way back to the 9th century, but it gained prominence only from 1189 onwards, when a Charter of the emperor Frederick I Barbarossa made it a Free Imperial City. The Charter granted Hamburg several privileges, including an exemption for its ships from the payment of duties on the Elbe River. In the first half of the 13th century, the town joined the Hanseatic League. This was an association of the mercantile towns of Northern Europe, long headed by Lübeck. Hamburg played a key role in the League. Even when the latter began to decline, the town continued to prosper, thanks to immigrant merchants from the Netherlands and France fleeing from religious persecutions.

In the first half of the 19th century, the South American states became important commercial partners of Hamburg's shipping companies and merchants. The harbour's cargo volume gradually increased so much as to require the construction of the city's first modern port basin, Sandtorhafen, built between 1863 and 1866 (Hamburg Port Authority, 2006; Hafencity Hamburg GmbH, 2009; http://www.hafencity.com). Over the course of the 19th century, the port evolved into the current HafenCity. Other port basins were built in the wake of Sandtorhafen – important, because of the transformation of the first modern port basin by the HafenCity redevelopment – namely Grasbrookhafen (1872–1881), Magdeburger Hafen (1872), Brooktorhafen (1880) and Baakenhafen (1887). To protect the port structures from flooding, the low marshes were gradually raised 4 to 5 metres above sea level. The Great Fire of 1842 destroyed a third of the old city, causing considerable damage to buildings. Reconstruction lasted over 40 years. In World War I, Hamburg lost many of its merchant ships (Figures 18.10 and 18.11).

In World War II, being at once a major public transportation node and an important industrial location, it became a target for Allied bombings that destroyed much of the port. After 1945, reconstruction began, and, concomitantly, the port experienced a significant growth in freight traffic.

*Figure 18.10* Hamburg, detail of the historic centre

*Figure 18.11* Hamburg, detail of the historic centre

Traces of its history are present in different places of the HafenCity area. The Speicherstadt warehouse plays an important role acting as an entrance and a joining element. Also due to the impressive brick-built facade – which remains almost visually unchanged – the old warehouse is listed as a historic monument and is designed to be acknowledged as a UNESCO World Heritage Site in 2014. The Speicherstadt has been now used for different purposes: museums and traditional goods storage, together with new functions were established, meant as a bridge between the old and the renovated image. These include those involved in the art and culture sectors as well as multimedia agencies and creative business.

The old harbour basins with their buildings represent for HafenCity another important element in recovering its place identity. The quay walls and the tracts of water constitute a place of attraction for both locals and visitors. The historical buildings which were retained, including the Speicherstadt as a whole, the emblematic Elbphilharmonie Concert Hall on the rooftop of Kaispeicher A, the International Maritime Museum in Kaispeicher B, and the old Port Authority building, became the landmarks of HafenCity.

The tangible identity has been enhanced together with the immaterial heritage of traditions which is being revived and reinterpreted in various ways, as in the case of the Sandtorhafen basin, where traditional steamships, sailing ships and cranes have been reconstructed.

The HafenCity project was approved in 1998 after an idea competition. The aim of the project is to allow the city centre to expand and, at the same time, integrate with the port beyond the Elbe, which for decades has been increasingly relegated to the southern-most part of the town.

The implementation of the master plan for HafenCity was preceded by a declaration of the strategic interest of the waterfront for the development of the Hamburg community. To complement the master plan, a number of detail plans were drawn up, including a transport plan, a land use plan, a plan for green areas and an evacuation plan in case of flooding.

The building of HafenCity has increased the surface of the original medieval city centre by 40% in 20 years. The 155 hectares of the medieval city's overall area are constituted by one third water and two thirds land. The whole operation is coordinated by a limited liability private company, HafenCity Hamburg GmbH, owned by the Free Hanseatic City of Hamburg, which manages relations between the public and the private sector.

The project is subdivided into eight main areas with an elaborate articulation of private and public spaces. Mixed use is achieved by devoting 3% of the available land area to advertisement, 8% to culture, science and education, 33% to habitation (http://www.hafencity.com; about 5500 apartments for 12,000 people are planned) and 56% to services, tertiary and tourism. Thirty-five per cent of the surface has been occupied by buildings, 25% by streets and transportation infrastructure and 36% by public spaces and private spaces accessible to the public. Only 4% of private spaces is not accessible to the public. Access to private spaces, especially where the most friction may occur between visitors, residents and people who work in the area, has been defined by an agreement guaranteeing and regulating their use by the public. Except for the promenades, the streets and parks are between 7.5 and 8 metres above the water level. This has created a new topography for the area, that enhances its "port ambiance".

The starting point in the building of HafenCity was the Am Sandtorkai quarter, where the first residential and office buildings were erected. The choice of this neighbourhood was determined by its position between the Speicherstadt warehouse district and Sandtorhafen, the nineteenth-century port basin. The buildings erected here are isolated from one another to provide them with an exclusive view of the city centre and the harbour. The only structures extending beyond the waterfront and the partially roofed promenade are the moored barges. These new buildings, erected only a few blocks away from the brick warehouses – which still house the stores of Oriental carpet merchants, as they have ever since the nineteenth century – were planned to harmonize with the historical buildings of the Speicherstadt.

On the other side of the Sandtorhafen is the Dalmannkai quarter, with the Dalmannkai houses and Elbphilarmonie Concert Hall, the result of an industrial reconversion project carried out by Herzog & de Meuron. The structure, shaped as a boat sail, is planned to house a concert hall, a hotel, a conference centre, a delicatessen and several residential units.

The new Dalmannkai quarter is the most heterogeneous area of all of HafenCity and has become a meeting point for residents, workers and visitors. A broad range of residential units, all facing the water, but differing in size, location and architecture, answer different needs. They range from luxury apartments – including, for example, some designed by Philippe Stark – to middle-priced ones. This variety is the result of an offer policy based on small-scale competitions to keep planning quality high.

Sandtorpark and Sandtorhafen, situated between Sandtorhafen and Überseequartier, are an integral part of the plan for the open spaces west of HafenCity. These were planned by the EMBT studio in Barcelona with materials and features matching those of the surrounding spaces and buildings, contributing to the project's sustainability.

The EMBT studio also planned Sandtopark two squares, the Magellan Terraces at the end of Sandtorhafen and the Marco Polo Terraces at the end of Grasbrookhafen. Here promenades and pedestrian paths run at different heights, alternating with green spaces and open spaces providing access to the wharves and the water.

On the south side of the area, at Elbtorquartier, is building Kaispeicher B, the oldest warehouse of HafenCity, now converted into a museum of navigation. Furthermore, a university for the study of architecture and the development of metropolises, the HafenCity Universität für Baukunst und Metropolenentwicklung, was realized on the banks of the Elbe running along the quarter.

The core of HafenCity is the Überseequartier. This is a mixed-function neighbourhood extending over 7.9 hectares, currently still in construction. A projected 40,000 visitors will eventually come here daily through the new cruise terminal located on the neighbourhood's waterfront.

The Überseequartier extends parallel to Magdeburger Hafen. The offer of the Überseequarter is enhanced by the Hamburg Maritime Centre, planned by the OMA studio, and the Hamburg Cruise Centre, designed by Massimiliano and Doriana Fuksas, a complex including a luxury hotel, a terminal for cruise ships and another for buses.

Another building with a strong visual impact is planned at the spot where the pedestrian zone reaches the Elbe. Here Rem Koolhaas built a Science Centre, a building shaped as a large vertical ring intended as a symbolic gateway visible from far away.

Am Lohsepark is a quarter on the east side of HafenCity – mostly completed in 2016 – connected to Brooktorhafen in the north by the renovated Ericus bridge, a former railroad bridge built in 1870. The quarter includes a mixed-use building for offices, health services, commercial space, social services, kindergartens and housing units. Among these, an inclusive household community, where people with disabilities and students live together is included. Hotels, restaurants and a park – the Lohsepark – that is the largest contiguous green space of the whole of HafenCity. The area – fully barrier-free – is of 4.4 hectares, of which 550 metres in length and 100 metres in width with more than 500 trees, paths and movable chairs to invite people to stop and take a break (https://www.hafencity.com/quartiere/am-lohsepark).
Finally, with respect to the mobility, in HafenCity the motorized individual transport was reduced from the 47% of Hamburg to the 20%. The HafenCity area lies less than a kilometre away from Hamburg City Hall and the city centre. It is thus easily reachable even on foot. Its services are readily available to the citizens of the rest of Hamburg as well as HafenCity's own residents. Seventy per cent of foot and cycling paths run from lanes which are separated by motorized traffic. The public transport, including ecological buses and river taxis and the new U4 subway stations, connects HafenCity with the rest of the city and contributes to the ecological, economic and socially sustainable integration of the new district. The new 10 kilometres of waterfront along the Elbe are thus truly reconnecting the city to its river (Porfyriou, Sepe, 2017; Hafencity Hamburg GmbH, 2004, 2006, 2022).

## The liveability and healthiness factors

The fifth phase concerns the identification of elements that contribute to the health and liveability of that place.

HafenCity is an area which has many elements of interest. In this study, we are referred to the public spaces which constitute an important part of this place because these favour its liveability.

Due to its nearness to the historical centre of Hamburg, but also thanks to the wild operation of participation which has been carried out from the

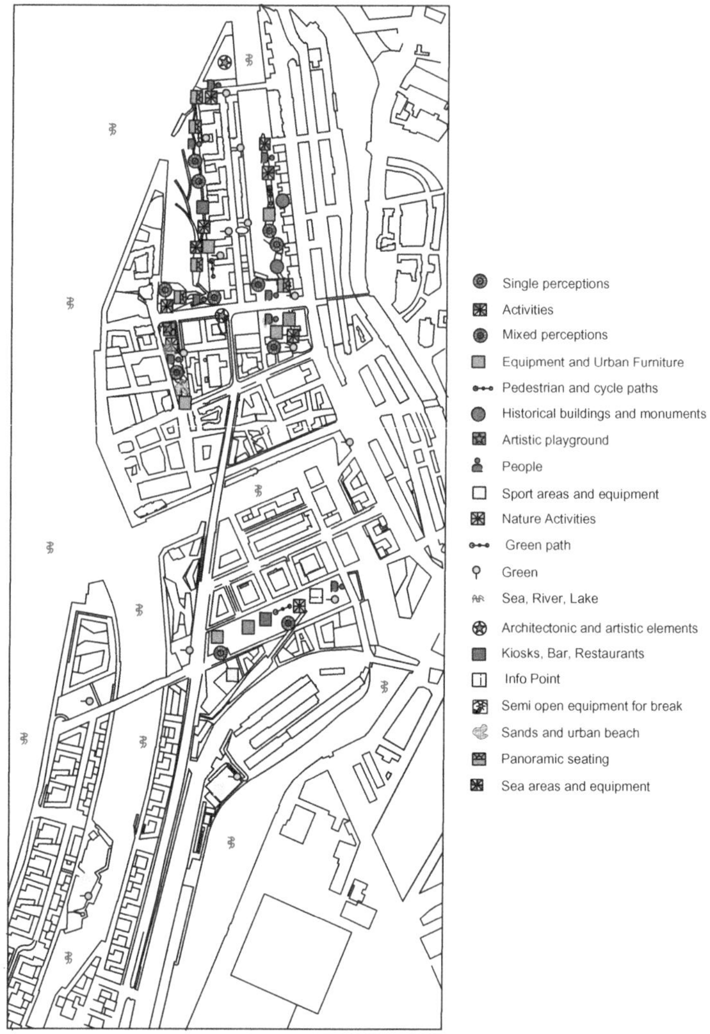


*Figure 18.12* Hamburg, HafenCity, map of urban health and liveability analysis

beginning of the project of regeneration, the area is well known to both inhabitants of the city and tourists and is well used (Figure 18.12).

The presence of public spaces with many possibilities of sittings has determined that both professionals who work there and locals who come from other areas of Hamburg use this place with frequency and at a low or moderate pace.

The presence of tradition, given by both the Speicherstadt warehouses and the historical sailing boats, and the new project interventions, given by the new architectures, offer an atmosphere of memory and innovation which people visiting the place like. The presence of an InfoPavilion and a little maritime museum on the Magellano Terraces fully inform people of the history of the place and the ongoing project of regeneration, offering a further element of participation.

Other aspects of success include different levels of surface which create dynamism and diversified uses, many visual perceptions and urban landscapes in perspective; an articulated design which creates variety for those who live in these spaces and an attractive place to walk and go by bike.

The realization of a wide playground for children – also used by students for educational trips – is a further aspect of success of the area, together with the presence of the new HafenCity University which creates a continuous use of the place. The sea is lived in its multiple "shape": as a background for architecture, as a landscape to be observed from different points of view and as a place to be lived and admired, thanks to the possibility of taking both public transport boats and touristic ships for different kinds of tours. The presence of bars and restaurants offers the possibility of extending the stop in a pleasant way. Serenity, joy, happiness, healthiness and surprise are the most solicited sensations in that place.

## Designing health and liveability

Checking the degree of healthiness, liveability and happiness is carried out through the 25 principles of urban health and liveable design concerning the Charter. By the check of the principles, those which could be improved include (2) to encourage the use of the place by people of different age groups, from children to older adults; (7) to create suitable spaces for dogs and domestic animals; (10) to improve suitable cycle lines; (15) to have the possibility of using the space in different weather conditions and seasons, contributing at the same time to its good state of maintenance; (17) to both allow and promote different types of functions such as games, breaks, walking and so on; (18) to facilitate gymnastic activities – also slow – with the presence of small equipment or a designated space; (20) to encourage the presence of art in its different forms; (22) to promote participation, namely the feeling of being able to contribute to the life of that place increasing the sense of belonging; and (24) to promote the educational function which a place has – for example clearly displayed information about history of the place and so on or in suitable ways to use it – increasing its intrinsic value.

The items of the questionnaire which follow derive from the principles which could be better satisfied:

1. What about creating a bike-dedicated lane?
2. What about enhancing or creating connections?
3. What about differentiating activities for different age groups?
4. What do you think about inserting artworks?
5. What about identifying sites for participation events?
6. What about inserting light structures for different climatic conditions?
7. What do you think about adding equipment for gymnastic exercises?
8. What about adding an area for dogs?

9. What do you think about adding information about the place?

About 140 people, including locals and visitors from all over the world, aged between 20 and 70 years old, were interviewed. The language used was English, and people generally answered with attention and pleasure. With respect to the first question, "What about creating a bike-dedicated lane?" 100% of those interviewed, both locals and visitors of all ages, answered that they agreed with this idea. Of these, 80% of locals added that the bike lane could be realized in the whole HafenCity area with related signage.

Regarding the question, "What about enhancing or creating connections?" 80% of locals and visitors aged between 30 and 70 answered that they agreed, adding that the connections should be realized with cycle or walk paths. The remaining 20% answered that they agree with this idea, adding that the connections could be realized also with green or light paths. All the interviewed people aged between 20 and 30 answered that the connections could be realized in different ways, namely with physical paths and with virtual ones (Figure 18.13).

With respect to the question, "What about differentiating activities for different age groups?" 100% of locals and visitors of all ages answered that they agreed with this idea. Locals added that this is important to also allow the young, older adults and people with disabilities to further enjoy the whole HafenCity area with dedicated game areas, paths, observation places and so on.

Regarding the question, "What do you think about inserting artworks?" 80% of interviewed people – both locals and visitors of all ages – answered that they agreed. In particular, locals added that artworks could be of different types, including sculptures, murals and light installations. The remaining 20% of people answered that they agreed, adding that the art products could be realized with sustainable materials and be both permanent and temporary.

Concerning the question, "What about identifying sites for participation events?" 70% of locals and visitors of all ages answered that they agreed with this idea; of these, locals, in particular, added that public space could be suitable for this function. The remaining 30% answered that they agreed and that participation events could be realized in different places according to these aims.

Regarding the question, "What about inserting light structures for different climatic conditions?" 100% of those interviewed – both locals and visitors of all ages – answered that they agreed with this idea. Of these, locals added that these structures could be useful above all in the low temperatures and rainy periods in which open-air spaces are less used.

With respect to the question, "What do you think about adding equipment for gymnastic exercises?" 70% of locals and visitors of all ages agreed with this idea. Locals added that the equipment should be for all and useable in all periods of the year. The remaining 30% of people answered that they

*Figure 18.13* Hamburg, HafenCity, map of urban health and liveability design

agreed and that this equipment should be inserted close to the access areas to HafenCity or the parking areas.

Regarding the question, "What about adding an area for dogs?" 100% of locals and visitors of all ages answered that they agreed. Locals added that off-leash areas are important to allow dogs and their owners to stay in the dedicated areas freely and, if these are equipped with dog games, to fully enjoy HafenCity.

Concerning the question, "What do you think about adding information about the place?" 100% of those interviewed agreed with this idea, adding that, although there is already an information centre, they would like information within the whole area. Of these, 40% of visitors added that the information could be also inserted in QR codes or apps.

In researching reviews on websites, on Instagram, the official page #hafencity_hh has 1,881 followers and 81 posts, #hafencityhamburg has 45,593 posts, another #hafencityhamburg has 804 posts and #hafencity has 491,308 posts. Furthermore, there are hashtags using the names of the public spaces, although there are few posts. Regarding Facebook, there is a non-official page – HafenCity – with 52,821 followers. On Twitter there are many pages: Hafencity Hamburg, @Hafencity_HH, with 889 followers; Hamburg Hafencity @HH_Hafencity with 113 followers; and Hafencity @ Hafencity_Watch, with 1,716 followers.

Considering booking pages, on TripAdvisor, present are HafenCity, with 1099 reviews; Speicherstadt – World Heritage Site, with 4,990 reviews; the Port of Hamburg, with 7,234 reviews; the Elbphilharmonie, with 3,636 reviews; and the International Maritime Museum, with 1,240 reviews. Aspects which are evidenced regard the operation of urban regeneration which is for them successful, the beauty of buildings and public spaces, the presence of bars and restaurants and the cleanliness of the area. Aspects connected to the relation with the sea, the old sailing ships and the boat for the visit of the area from the sea are also evidenced with positive comments.

From the information gathered in the previous phases of the project, the following project interventions resulted.

The first project intervention is to create a bike-dedicated lane. Public spaces in HafenCity are many, and it is possible to cycle freely. To improve the health of this place, a dedicated and continuous cycle path with related signage could be added; this could invite more people to practice physical activity, enjoy the beautiful sightseeing that all the area offers and work as a connector between the different public spaces.

The second project intervention is to enhance or create connections. As emerged from the surveys and questionnaire, public spaces are of different kinds and situated in different parts of HafenCity. To further improve the health, liveability and happiness of this place, the creation of a connection between all of them through a dedicated cycle lane – before mentioned – a walk path or a green path or other modalities could be carried out – such as virtual paths – that should also be related to the historical centre.

The third project intervention is to differentiate activities for different age groups. People of different ages have different needs and ways to enjoy a public space when they go. To increase the health, liveability and happiness of HafenCity, games, sport and educational activities could be added to allow not only children and adults but also young people, older adults and people with different abilities to fully use this place.

The fourth project intervention is to insert artworks. Art in its different forms represents an important factor for liveability and happiness of a place.

To improve these elements in HafenCity, further art products – both permanent or temporary – such as sculptures, artistic furniture, murals and light installations could be added to the public spaces.

The fifth project intervention is to identify sites for participation events. HafenCity is a wide area, and many places could be used for participation events devoted to different themes and to residents, locals and visitors. Public spaces – such as Lohsepark and Grasbrookpark for example – could be the best for this function thanks to their wide extension and space for free activities.

The sixth project intervention is to insert light structures for different climatic conditions. In the periods in which the weather conditions present too high or too low temperature or rain or snow, light and mobile structures could be inserted to protect people from unpleasant climatic conditions and allow them to experience HafenCity in all the periods of the year.

The seventh project intervention is to add equipment for gymnastic exercises. HafenCity offers many possibilities of leisure and relaxation, and the insertion of equipment for gymnastic exercises – in Lohsepark, Grasbrookpark or the other public spaces – could furtherly improve its healthiness. The equipment should be designed to be used by people of all age and ability and made by natural materials.

The eighth project intervention is to add an area for dogs. Off-leash areas could be added in Lohsepark and other public spaces of HafenCity to allow dogs and their owners to enjoy the place freely; dedicated games could be added to further improve the liveability of those areas.

The ninth project intervention is to add information about the place. HafenCity is a wide area with squares, promenades and parks of different kinds and typologies of activities to carry out. To further improve the health and liveability of this place, information points – also with the use of new technologies – could be added. These could be in continuity with the information already available in the HafenCity Information Centre and guide the visitor to fully enjoy the place.

## Conclusion

The HafenCity regeneration project, started in the late 1990s, has restored the maritime vocation of the historical city and is strictly connected with the city. The area is emblematic as a case study because this is a new area in which a new urban identity was created for the whole city, improving its liveability, healthiness and degree of satisfaction of both locals and visitors.

The listening process activated from the early stages of the design operation and the idea of recreating the maritime identity were very important elements for the success of this place. The case study analysed the public spaces that cross HafenCity and, in particular, Grasbrook, Marco Polo Terraces, Dalmannkai Promenade, Vasco de Gama Platz, Magellan Terraces, Überseeboulevard and Lohsepark.

From the surveys and questionnaire during the phase of analysis, it was observed that these public spaces – for their variety and articulation – offer multiple possibilities of activities for different surface subjects with different needs and preferences.

The aspects of success include different levels, many visual perceptions and urban landscapes in perspective; the historical memory given by the Speicherstadt warehouses and the sailing ships visible from the different points of views; an articulated design which creates variety in who live it; and an attractive place to take a break, observe, walk and bike.

The playground for children is a further aspect of success of the area, together with the presence of the new HafenCity University which creates a continuous use of the place. Furthermore, the sea is perceived not only as a landscape to be observed from different points of view but also as a place to be lived and admired.

This area is also very close to the historical centre of the city and is easily reachable with different means of transport, creating facility of access and crossing.

Through the phases of design, the project interventions identified include creating a bike-dedicated lane, enhancing or creating connections, differentiating activities for different age groups, inserting artworks, identifying sites for participation events, inserting light structures for different climatic conditions, adding equipment for gymnastic exercises, adding an area for dogs and adding information about the place.

Indeed, the multiplicity of public spaces could be further improved regarding their liveability and healthiness by creating dedicated paths for bikes and pedestrians which connect them. Another important aspect is constituted by the weather as well, which, in particular, during the winter is frigid. The frequency of public spaces in these periods is more limited for questions connected to the weather, although the presence of public buildings and private societies offers many possibilities of use the place during the whole year. Accordingly, as aforementioned light and mobile structures could be added to protect people from unpleasant climatic conditions. A last "ingredient" which is already present in different ways, both in the current spaces and in the proposed project interventions, is flexibility that, together to the attention to the respect of the old and new place identity, can assure the success of this area in the next years as well.

# 19 Madrid

## The area

Madrid, the capital of Spain and its most populous city, is crossed by the Manzanares River and is characterized by the presence of many parks and gardens, including the contemporary Madrid Rio, the object of this study.

The on-site visits consist in the observation of the places in order to decide what parts include or not in the study area. In the case of the Madrid Rio Park, due to its urban continuity and long extension, it was decided to focus the attention on the park itself in its entireness.

Madrid Rio Park is limited by the Puente del Rey in the north-west part and by the Puente de La Princesa in the north-east part and is extended long the Manzanares River.

Observations were carried out both during the weekly and week-end days and both in the morning and afternoon hours. The area was divided in two parts for the surveys and questionnaires, namely Puente del Rey–Puente de Arganzuela and Puente de Arganzuela–Puente de La Princesa. Illustrations of the results are reported, starting from the Puente del Rey–Puente de Arganzuela part.

## People and activities

The first survey concerns the kind of people which use this place. The majority of people is constituted by locals, residents and professionals of all ages. The presence of many activities attracts both young people, adults and older adults. Many tourists visit this park as well, both for sport activities and for taking a break.

The second survey concerns the activities and relative frequency. The main activities include walking, running and cycling, in particular and, walking in the paths, sitting on benches, working for garden maintenance, taking a break sitting outside a kiosk, eating in a restaurant, working in the bar/restaurant, skating and taking a break. The walking activity is carried out with high frequency by people – locals, residents, professionals and visitors – of different ages during all the hours of the day. Running, cycling and skating

DOI: 10.4324/9781003098775-24

activities are carried out mainly by people both young and middle-aged – the latter mainly with regards to running and cycling ones – in morning and afternoon hours with medium frequency. Sitting on benches and taking a break are activities for all ages including mothers with babies and older adults. These activities are carried out with high frequency and both in the morning and afternoon hours. Taking a break sitting outside a kiosk and eating in a restaurant are activities which are carried out with a medium-high frequency by people of age both locals – including professionals – and visitors, depending on the hours. The high frequency is in general during lunch and dinnertime. Working for garden maintenance is carried out by gardeners with high frequency due to the great extension of the park and the green areas which require continuous maintenance. Crossing the bridges is another very frequent action for the same reason: the great extension of the park and the presence of many connections between the west and east sides of the site.

Furthermore, observation by the bridges and the viewpoints are actions which are present in Madrid Rio and are carried out in the morning and afternoon hours by all people with a medium frequency together with the action of taking photos, the latter carried out mainly by locals and visitors.

Playing with the "tirolina", playing tennis, doing gymnastics using the equipment of the park and meeting for doing together open-air gymnastic are actions which are carried out with a medium frequency by people of all age and mainly locals in the morning and afternoon hours. In particular, the activity of meeting to do open-air gymnastic together is carried out by older adults above all in the morning and in a dedicated space. Playing in the playgrounds is carried out by children of different ages and by children with disabilities with medium frequency.

Meeting of classroom is an activity that is carried out by teachers and students all together with medium frequency above all in the morning hours, for different kinds of lessons or visits.

Taking the children in the stroller, taking the dog for a walk and bringing people in wheelchairs are activities which are carried out with medium frequency by people of different age, mainly locals.

Going with roller, going on electric scooter, playing "petanque", taking the sun on the grass and drinking water by the fountains are that are activities that are carried out in the park with lower frequency with respect to the other by locals, professionals and visitors. Finally, other activities include experiencing the smells of nature and reading the information point carried out by all people with a medium-low frequency.

Then third survey of single and mixed perceptions was carried out. The single perceptions are all involved and pleasant. The visual ones concern the different flora of gardens; the orchard; the church; the line of trees; the Puente del Rey, the Puente de Segovia and the Puente de Segovia's gardens; the Puente obliquo; Parque de la Virgen del Puerto; the different shadows and perspectives created by the Puente Verde en Y and the Arganzuela bridge; the residential buildings from the streets which border the park; Vicente Calderón Stadium; the playground areas for children; the wood games for children; the Puente de Toledo; the long wood benches (choreographic); and the river with its vegetation and fauna.

Each bridge has a different conformation and produces a singular relationship with the urban context, the paths and the nature creating particular visual perspectives and shadow effects. Furthermore, some bridges in the under part are used for gymnastics and present drawings or graffiti.

The tactile perceptions include sand with little stone, the sands of the "playa", grass, gravel floors, wood games for children and wood benches.

The acoustic perceptions include work machines of gardeners, the water of fountain, the voices of people, the voices of children who are playing, the sounds of music used by people who practice open-air dance or gymnastics, the rope of the "tirolina", the cars crossing the vehicular bridges and young people who play on skates or roller blades. The smell perceptions include flowers, trees and nature in general, the river, and the grass.

The mixed perceptions are mainly of serenity, joy, happiness, urban liveability and healthiness. In some places, namely those less used such as the parts with the fountains, the mixed perceptions include the calm and the sensation that the time is going slower. In other parts, such as the Arganzuela footbridge, the perception is of surprise because of the particularity of the architecture. In the skating area, the perception is of vivacity due to the acrobatic performances of younger people. In the open-air fitness area for older adults, the perception is of wellness.

Then the last survey is that of elements which contribute or less to the perception of healthy and liveable places. These include playgrounds of different kinds, shape and materials – with sands, wood, aluminium and so on, all of good quality and many choreographic – and many for those who are disabled; benches of different materials and sizes – some are very long; gymnastic equipment in many parts of Madrid Rio; sport features for basketball, skateboard, football, "petanque" and tennis; gardens, orchards and little slopes; cycle paths; pedestrian paths; many access areas, ramps and steps; fountains; the Manzanares River in regeneration; information points; information on pavements – concerning the kilometre of the park or if a path is for cycles or pedestrians; many bridges and transversal crosses; fountains with drinkable water; sculptures; kiosks with foods; a beach; deck chairs; points of view; and picnic areas.

## The places from peoples' points of view

The questionnaire administered to people who use the place is the third phase of the method. This was administered to 120 people – residents aged between 25 and 65 years – and visitors from Europe and US aged between 35 and 70 years. The languages used were English and Spanish, and people were generally willing to give answers to the questions.

Questions included the following:

1. Does this place give you a feeling of happiness or sadness/liveability or discomfort/healthiness or unhealthiness?
2. What are the elements that give you the above sensations?
3. What are the main facilities that give quality to this place?
4. What are the activities that you do in this place, and how often?
5. What do you think about the presence of many or few people here? Do you think that it is capable to improve the pleasantness or unpleasantness of the place?
6. According to the place's healthiness, liveability and happiness, what could be done in order to improve this place?
7. What is a healthy/happy/liveable place that you remember in this city or elsewhere?
8. How the weather condition might influence the perception of healthy, liveability or happiness of this place?

In the following, questions and answers are reported.

To the question, "Does this place give you a feeling of happiness or sadness/liveability or discomfort/healthiness or unhealthiness?" 100% of people gave a positive answer independent from nationality and age. Both locals and residents added that to think that this area was a highway and now was a big public space is surprising. For both visitors and inhabitants, the place gave them a perception of liveability, quietness and happiness.

To the question, "What are the elements that give you the above sensations?" 60% of locals, independent of age, answered the rose garden and the other gardens and orchards, the river and the playgrounds, while the remaining 30% answered the bigness of the park and the presence of many paths surrounded by greenery. Ninety per cent of visitors answered the general atmosphere, with no specific details concerning the elements which contribute to the sensation of liveability. The remaining 10% of them gave some more information referring to both the historical and new bridges and to the shape of gardens.

The answers concerning "What are the main facilities that give quality to this place?" were different. Of locals, 60%, independent of the age, affirmed that the main facilities are constituted by the many benches spread in the park and the many accesses. For the other 40%, young people and adults answered the running, walking paths, the sport areas – gymnastic, tennis, basketball, football, skating and so on – and the mothers with children answered the playgrounds accessible to all – also to people with disabilities – and the paving suitable to people pushing children strollers and others in wheelchairs. Visitors answered that they like the possibility to arrive there with public transport, the benches, the kiosks. Furthermore, all the respondents added that there are many information points indicating the route and other information.

To the question, "What are the activities that you do in this place and how often?" 65% of locals – both young people and adults – answered that they frequently go to walk and take a break while observing the gardens or eating/

drinking something. The other 35% of locals answered that they frequently bring children to the playground, run, go by bicycle, play some sports and meet other people. The 80% of visitors answered that the main activities that they do in Madrid Rio include walking and sitting to take a break. The other 20% act open-air sports such as running and going by bicycle.

As regards the question, "What do you think about the presence of many or few people here? Do you think that it is capable to improve the pleasantness or unpleasantness of the place?" about 100% of people – both locals and visitors of all ages – observed that the park has a considerable extension, and for this reason, the concentration of people usually is not high.

To the question, "According to the place healthiness, liveability and happiness what could be done in order to improve this place?" 90% of interviewees answered nothing. The remaining 10% of people answered that they would like a major presence of water – scenic fountains and drinking water fountains.

Regarding "What is a healthy/happy/liveable place that you remember in this city or elsewhere?" 70% of locals answered the Parque de El Retiro in Madrid, although it has different characteristics from Madrid Rio, while the other 30% answered that this park is unique in Madrid. Visitors answered many different parks or regeneration areas including Central Park in New York, Green Park in London, the waterfront of Berna and many parks of Paris.

To the question, "How the weather condition might influence the perception of liveability or happiness this place?" 100% of visitors answered that in general the weather can change the perception of a place, but they do not know how the weather is in the park during the different seasons, so they are not really capable to give an answer. All the locals answered that usually the weather in Madrid is quite good, and, for this reason, the perception of this park can only change during the winter season.

## The transformation of the area

The traditional analysis has given as a result that Madrid Rio is a 10-kilometre linear park, designed by the West 8 and MRIO (Burgos et al., 2014) architects and completed in 2011 which is situated along the Manzanares River in Madrid in the area that was before covered by the M30 highway (Abajo et al., 2020). The river is not located in the city centre but is connected to it by many urban axes and streets.

On west and east parts of the park are both residential and commercial buildings and the football stadium. Many accesses along the areas are designed with metro stops close to them. The shape of the park is curved, and many bridges crosses the area including the Puente del Rey, Puente de Segovia, Puente de Toledo, Puente oblicuo, Puente Verde en Y, Puente Invernadero and Matadero and the Arganzuela bridge. Puente del Rey was built in 1816 during Ferdinando VII's reign and is part of a group of monuments including Puerta del Rey gate. Puente de Toledo was realized in the middle of the 1600 by Juan Gomez de Mora and commissioned by Felipe IV. In 1715, it was rebuilt by Pedro de Ribera, and in 1735, two new sculptures of San Isidoro and Santa Maria de la Capeza were added. Puente de Segovia, which is the oldest of Madrid, was built between 1572 and 1588 by Juan de Herrera and commissioned by Felipe II. Puente oblicuo is a modern bridge – similar to a promenade – which was built with the M30 highway and used for road traffic, that is now a 150-metre path with grass and trees inaugurated together with the park in 2008. Puente Verde en Y is one of the new bridge of Madrid Rio and is built in steel with two arms, on the left and right sides. The Puente de Invernadero and Matadero are two new bridges which connect the river with the Greenhouse and the Matadero Madrid cultural centre. These twin bridges were designed by West 8 and MRIO architects and are decorated with mosaics that represent 50 Madrid neighbourhoods.

The other new bridge is the Arganzuela one, built by Dominique Perrault, with a spectacular metal spiral shape. This presents two parts which connect the two banks of the river and is for pedestrians and cyclists but has also a direct access to the park below (https://www.madrid.es; Hernández, 2011).

The other monuments of the park include the Ermita Virgen de Puerto Chapel built in 1715 by Pedro de Ribera and rebuilt after the damages suffered during the Civil War period and the Matadero and the Mercado de Ganados built in 1931 by Luis Bellido and now recovered as a contemporary art centre.

The map presents walkways, cycling routes, green cyclist ring, many gardens and orchards – with banana, chestnut, magnolia, cedar, lime, oak trees – a beach, playgrounds for children and sport facilities and fitness areas. The shape of the gardens – mainly curved – changes according with the different part of the map but maintaining a general harmony with the context and a perception of continuity of the design.

The important part of the area is the river, in the course of renaturalization, and its ecological corridor. The objective of the regeneration includes the connection of all the river sections of the city, enhancing both the beauty of the landscape – and its flora and fauna – and social value of it in the urban context.

## The liveability and healthiness factors

The last part of the analysis concerns the identification of places and activities which make the site liveable, healthy and happy from a user's point of view.

Madrid Rio is an area of regeneration which is extended for 10 kilometers. The main characteristics which make it liveable and healthy mainly concern the right mix among the many and different elements which compose it and the good quality of materials and design. There, almost all the needs for an open-air space can be satisfied.

The first element is the continuous paths which cross all the area. These paths are conceived for many activities including walking, crossing, cycling and running on a pavement that is of good quality and comfortable. These contain different possibilities of break and stop dedicated to different necessities; playgrounds built with different materials, shapes and features and for this reason devoted to all ages; playground for those who are disabled, with games suitable for people with different abilities; and spaces for fitness activities. These activities are many and are spread in all the park, easy to access, always opened and surrounded by green. Some of them have not only the function of game but also have a good and scenic design, well integrated within the context. Then, there are sport structures for tennis, paddle ball, football, rock climbing, skateboarding, roller skating, BMX biking, "petanque", basketball and structures for all the sports needs of residents, locals and visitors. The fitness areas – for all age and ability –are disseminated along the park giving to people the perception of a healthy park. Also, these structures are well inserted in the general design of the park. The sport activities are integrated with the green which is present in many shapes in Madrid Rio. Gardens and orchards present shapes, colors, smells and functional organization – paths, benches, and illumination – which invite to admire, stop or observe them from above (Figure 19.1–19.13).

Historical monuments and bridges characterize the park together to the new ones including, Puente Verde en Y, Puente Invernadero and Matadero and the Arganzuela ones. Every bridge generates a particular relationship with the surrounding site, creating a large variety of landscapes – games of shadows, sightseeing and places where people like to cross and take photos, selfies and so on. The latter, in particular, has a scenic shape that is very attractive and create a suggestive imagine which is often used as a background for photographic artistic set.

Furthermore, the bridges allow many accesses to Madrid Rio which, together with stairs and ramps, constitute an important element for the life of the park. People can entre here very easily and from many points of the city.

*Figure 19.1* Madrid, Madrid Rio Park perspective

*Figure 19.2* Madrid, Madrid Rio Park, Puente de Toledo

*Figure 19.3* Madrid, Madrid Rio Park, detail of a path

*Figure 19.4* Madrid, Madrid Rio Park, Arganzuela footbridge

*Figure 19.5* Madrid, Madrid Rio Park, detail of a playground

*Figure 19.6* Madrid, Madrid Rio Park, detail of skate area

*Figure 19.7* Madrid, Madrid Rio Park, Puente de Invernadero

*Figure 19.8* Madrid, Madrid Rio Park, detail of a path

*Figure 19.9* Madrid, Madrid Rio Park, detail of a playground

*Figure 19.10* Madrid, Madrid Rio Park, detail of a sport area under the bridge

*Figure 19.11* Madrid, Madrid Rio Park, detail of an area for physical activities meeting

*Figure 19.12* Madrid, Madrid Rio Park, detail of a path with a sitting area

*Figure 19.13* Madrid, Madrid Rio Park, view of the Matadero

The large meadows have the double function of natural elements and places to sit, relax, take the sun, read a book, write, draw and observe. Not least, these have an agreeable smell.

In Madrid Rio, there is also an urban beach with a large fountain, sand and beach furniture which is a place to stay and socialize mainly in good weather periods.

Many other fountains are collocated in the park, although not constituting the main attraction of the place. The river, instead, represents the protagonist of Madrid Rio, with its sober beauty linked to its flora and fauna natural elements. People like to stop and observe the course of water, the gardens and the fluidity of the natural elements.

Finally, Matadero Madrid is an old industry recovered for cultural functions which constitutes a further element to have a full use of the area (Figure 19.14).

## Designing health and liveability

Checking the degree of healthy, liveability and happiness is carried out through the 25 principles of urban health and liveable design concerning the Charter.

By the check of the principles, Madrid Rio is in line with almost all. In all its extension, as it has emerged by the previous phases, the park contains a wide variety of elements which makes it healthy, liveable and happy.

The principles which could be enhanced include (7) to create suitable spaces for dogs and domestic animals; (15) to have the possibility of using the space in different weather conditions and seasons, contributing at the same time to its good state of maintenance; (20) to encourage the presence of art in its different forms; (21) to promote sculptures, games, or other elements and amenities which can bring a smile to a person's face promotes a state of liveability and happiness; (22) to promote participation, namely the feeling of being able to contribute to the life of that place increasing the sense of belonging; and (25) to facilitate the use of new technology to increase the knowledge of its intangible values and history, offering a more profound experience of the place.

The items of questionnaire which follow derive from the principles which could be better satisfied.

1. What do you think about creating some light urban structures to use in the park during cold periods?
2. Water is an important element of the park. What do you think about further enhancing it with more fountains or water games?
3. The Manzanares is the protagonist of Madrid Rio. What about creating an exhibition along the park to narrate in a wider way the history, including that of Madrid Rio?
4. What about adding artworks along Madrid Rio, including sculptures, to improve its attractiveness?

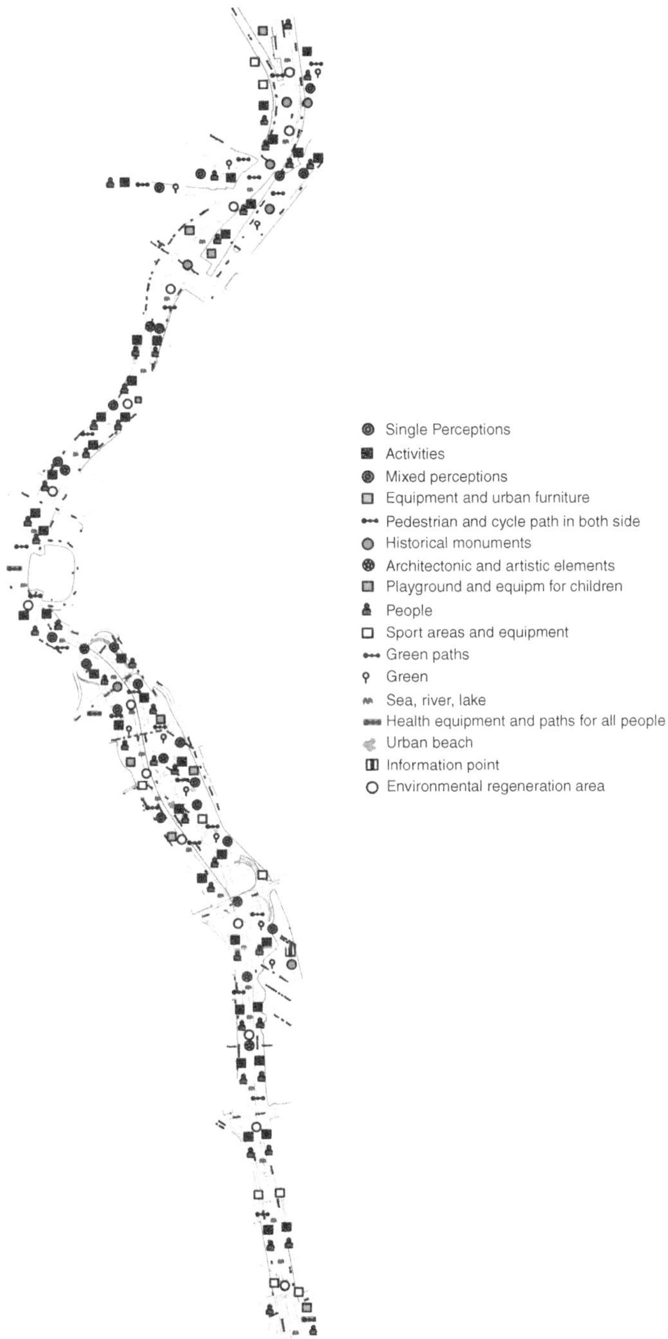


*Figure 19.14* Madrid, Madrid Rio Park, map of urban health and liveability analysis

5. Could participation events enhance the feeling of belonging of the park, making it more used by tourists?
6. Major technology companies have established in the area surrounding Madrid Rio. What about improving the use of new technologies in the park, such as adding QR codes and virtual maps?
7. The great extension of the park is an important element. What do you think about improving the rental of different means of transport, including electric scooters, bikes of many kinds, canoes for the river and so on, to favour its whole crossing also in dedicated spaces?
8. What about adding an area for dogs?

The questionnaire was administered to 100 people – residents aged between 25 and 65 – and visitors from Europe and US aged between 35 and 70. The languages used were English and Spanish, and people – as for the previous questionnaire – were generally very willing to give answers to the questions.

Regarding the first question, "What do you think about creating some light urban structures to use in the park during cold periods?" 100% of people gave a positive answer. On this question, about 20% of people added that the structures should be in line with the peculiarities and environment of the park.

With respect to the second question, "What do you think about further enhancing the presence of water with more fountains or water games?" 100% of people – both locals and foreigners – answered that they agree to do it in some areas because, although the Manzanares is already present, to have more fountains and water games is important both in the summer period and, more generally, as a factor capable of giving agreeable perceptions.

Regarding the third question, "What about creating a permanent exhibition along the park to narrate in a wider way the history, including that of Madrid Rio?" 100% of local people gave a positive answer. They remember the previous presence of the M30 and the big change due to the creation of this park. Although the environmental regeneration is explained in signage placed close to the Manzanares River, local people were positive about knowing more of the general history of the park, its bridges and the gardens. With respect to foreigners, 50% – aged between 45 and 70 years – were positive about this idea. The other 50% answered that they are quite interested in it but that it needed to be inserted using interactive totems, QR codes and so on.

With respect to the fourth question, "What about adding artworks along Madrid Rio, including sculptures, to improve its attractiveness?" different answers were given. Fifty per cent of both locals and foreigners answered that they agreed with this proposal without adding anything else. The other 50% – in particular locals between 45 and 65 years old – answered that they agreed with the idea, adding the importance of inserting artworks in line with the identity of this place.

Regarding the fifth question, could participation events enhance the feeling of belonging of the park and making it more used by tourists, the 100% of people generally agreed with this idea. Locals between 50 and 70 years added that the events should be organized regularly in the same place to favor an overall participation of people, while those aged between 35 and 50 added that the tool of communication is important for organizing participation events, which have to be, in particular, social networks. Of tourists, 100% agreed with this idea, adding that meetings should be connected with music, art and performance events.

The sixth question, "What about improving the use of new technologies in the park, such as adding QR codes and virtual maps?" 100% of people, both locals and tourists, agreed. In particular, people – both locals and tourists aged between 35 and 40 – added that technologies could be aimed at connecting different parts of the park and/or improving knowledge related to the different activities to do.

To the seventh question, "What do you think to improve and/or introduce the rental of different means of transport, including electric scooters, bikes of many kinds, canoes for the river and so on, to favor its whole crossing also in dedicated spaces?" 100% of those interviewed – both tourists and locals – agreed on slow mobility, introducing and improving the rental of different kinds of means of transport. About the 20% of people – locals aged between 40 and 60 – added the necessity of suitable signage to assure low speeds in case of more means of transport and pedestrians being present.

To the last question, "What about adding an area for dogs?" 100% of locals and visitors of all ages answered that they agreed with this idea. In particular, locals added that it is important to have some off-leash and game areas for dogs in a park.

Checking social networks and websites with suggestions, no official Instagram page is present, but many #madridrio hashtags are present, with 113,604 posts and 14,170 posts, among others. There is a Madrid Río Facebook page, with 5,053 followers, and many movies on YouTube.

Among the booking websites with suggestions, on TripAdvisor, there are 2,168 reviews, most of them with positive comments on the park which underline the surprise of seeing the river – and which before the park few people knew – its beauty and the many activities to do. Some comments include the non-proximity to the centre of Madrid but the presence of many metro stops, the great extension of the park which sometimes doesn't allow tourists to cross it wholly and the low water presence in the river during the warm periods.

Now the identification of project interventions will follow to improve the healthy, liveable and happy characteristics of Madrid Rio.

The first project intervention is to create urban structures to use the park in the cold and rain periods.

Madrid Rio is a park full of gardens and equipment activities. In order to improve its use during the cold and rain period, light urban structures could be added in line with the urban design and the architecture of the park.

The second project intervention is to enhance the presence of water and water games in their different shapes. Water is an important element in a park, and Madrid Rio is based on a river recovery. Although the presence of water in this park is wide, many fountains are in narrow areas and not always improve the beauty of whole park. Indeed, the fountains are often discovered casually. To enhance this factor, some fountains could be added along the park, and more scenic water games can further characterize those already existing.

The third project intervention is to realize a wider cultural/virtual/physical connection between Matadero and Madrid Rio. Matadero is a culture structure with many uses. The connection between the park is mainly realized through a bridge and a path but only in the final part of Madrid Rio. More possibilities of cultural/virtual/physical connections between Matadero and Madrid Rio could improve the use of both.

The fourth project intervention is to add artworks along Madrid Rio, including sculptures, to improve its attractiveness.

Art is a fundamental element in a public space. The insertion of some sculptures or installations could improve the liveability of this place, also creating a connection with the production of art of the Matadero.

The fifth project intervention is to create light exhibition areas and furniture along the park to narrate the history of both the park and the river.

The Manzanares is known by Madrid citizens but not by foreigners who usually do not connect the Madrid idea to that of city crossed by a river. This is because, differently by many cases, the river does not pass through the city centre. An exhibition which explains the history of the river and the park could help in this sense. This could be realized designing light exhibition panels to be put close to the river. The panels could be used for a temporary exhibition as well.

The sixth project intervention is to devote more areas to organizing participation events to enhance the feeling of belonging of the park by citizens and improve its use during colder and warmer seasons as well.

Participation events are important not only before the realization of an urban project but also after. The life of a place can change during the year according with new habits and needs. To improve the use of Madrid Rio, in particular in the colder and warmer seasons, events which involve residents and locals could be done also with educational functions.

The seventh project intervention is to improve the use of new technologies in the park adding QR codes and virtual maps. Near to the park there are many new technologies companies. Further uses of the park related to new technologies could be introduced, adding QR codes and virtual maps which could also help people with disabilities visit and fully experience the whole park.

The eighth project intervention is to improve the whole cross of the park, inserting the rent of different means of transport, including electric scooters, bikes of many kinds, canoes for the river and so on, and with dedicated paths, respectively.

The great extension of the park is a quality but not all people cross it in its entirely. The possibilities to use different means of transport which make it crossable, admiring the paths, gardens, orchards and the various areas could constitute an added value to its liveability and healthy.

The ninth project intervention is to create an area devoted to dogs. Dogs should have their own space to play with other dogs and together their owners. Madrid Rio has many suitable places which can be used for this function, giving the dog owners' the possibility to walk and play with them (Figure 19.15).

*Figure 19.15* Madrid, Madrid Rio Park, map of urban health and liveability design

## Conclusion

Madrid Rio is a new 10-kilometer park, the result of a wide operation of urban regeneration. It is a park of new generation because it was created to satisfy all the needs of people. People of all ages have the possibility to use the park thanks to many kinds of equipment, furniture and fitness structures.

Furthermore, the history of the bridges is well contextualized with the contemporary architecture of the new ones, from the spectacular Arganzuela Bridge drawn by Dominique Perrault to the Invernadero and Matadero ones with the beautiful murals, and the Puente Verde en Y.

Gardens and orchards created a very beautiful landscape around the Manzanares donating an atmosphere of harmony to all the place, satisfying the characteristics of urban healthy, liveability and happiness. As a result of the Healthy Pl@ce Design method, some elements could be added to improve these which were described in the previous section.

Madrid Rio is a sort of counterpoint to the other parks, such as El Ritiro, situated in the city centre. The possibility of reaching it by the metro and the bus through the many park entrances favors its use. Important elements are also the connection which it creates with many parts of Madrid and the fact that the park enhances the presence of the river which is unknown to many people from Madrid and outside.

This wide area constitutes a kind of a from the dynamism of a metropolis such as Madrid which has been in continuous change over the last years. The idea of innovation and tradition are here an important factor of success.

The identified project interventions include creating urban structures to use the park during cold and rainy periods; enhancing the presence of water and water games in their different shapes; realizing a wider physical cultural/virtual/physical between Matadero and Madrid Rio; adding artworks along Madrid Rio, including sculptures, to improve its attractivity; creating light exhibition areas and furniture along the park to narrate the history of both the park and the river; devoting more areas to organize participation events to enhance the feeling of belonging of the park by citizens and improve the use of it also in the colder and warmer seasons; improving the use of new technologies in the park by adding QR codes and virtual maps; improving the whole cross of the park, inserting rental options for different means of transport, including electric scooters, bikes of many kinds, canoes for the river and so on, and respectively dedicated paths; and creating an area devoted to dogs.

These project interventions for the improvement of the park have to be meant as a further enhancement of what already exists in terms of healthiness, liveability and urban happiness and the long sustainability of the area.

# 20 Newcastle–Gateshead

## The city

The case study concerns the quayside in Newcastle and nearby Gateshead city in England. Newcastle upon Tyne is a city in the county of Tyne and Wear, founded in the Roman period. Newcastle is located on the north bank of the River Tyne, the port of which developed in the 1500s and for a long time was one of the leading shipbuilding and repair centres in the world. Gateshead is situated on the southern bank of the River Tyne opposite Newcastle upon Tyne and is connected to it by many bridges.

In the 1970s, the industries suffered a major crisis and a subsequent closure of many of them, with a consequent economic decline (Comunian, 2009). The strategic actions for the redevelopment of abandoned areas in Newcastle concern the establishment of industries in innovative sectors, the construction of new homes and the lowering of rents of existing ones, the creation of digital services, the realization of a sustainable transport service and an increase in cultural and entertainment activities and events (Carta, 2007).

Since the early 1990s, culture has taken on an increasingly important role in urban policies until it becomes one of the main elements around which to transform the territory through public investments aimed at involving the population in activities related to art and culture.

## Urban planning strategies

The regeneration process of Newcastle, which also took place through collaboration with the nearby municipality of Gateshead, aimed at an urban change strongly linked to the local identity, paying attention to the participation and involvement of the population, the liveability of places, the long-term sustainability interventions and the creative economy (Florida, 2005; Carta, 2007).

The general strategy and urban core plan which were adopted in March 2015 form a strategic planning framework for development in Newcastle and Gateshead up to 2030.

DOI: 10.4324/9781003098775-25

The Vision for Gateshead and Newcastle upon Tyne 2010–2030 is based on a shared framework of general strategic policies and more site-specific policies. The realization of a common plan was due to the recognition of the advantage that both local authorities can have in terms of economic growth and trade, employment development, housing growth and infrastructure enhancement. The overall goal is to create two sustainable and attractive cities where people decide to live, work or visit them for their high quality and lifestyle.

Twelve strategic objectives were identified and included in the policies defined by the Plan for Gateshead and Newcastle upon Tyne 2010–2030. These include SO3, to increase competitiveness by improving and expanding the role of the urban core as the regional destination for business, shopping, education, tourism and as a place to live; SO5, to expand leisure, culture and tourism providing for all age groups and diversifying the evening economy; SO8, to improve sustainable access to, within and around the urban core by promoting fast and direct public transport links to the heart of the urban core, increasing walking and cycling and minimizing through traffic; SO9, to ensure the development and use of land protects, sustains and enhances the quality of the natural, built and historic environment, making the urban core a high-quality exemplar for Gateshead and Newcastle and ensuring communities are attractive, safe and sustainable; S010, to provide the opportunity for a high quality of life for everyone and enhance the well-being of people to reduce all inequalities; and SO12, to create a network of green public spaces connected and accessible by all.

The plan identifies various themes for strategic policies, of which people and place are the most transversal (Newcastle and Gateshead Council, 2015).

With respect to placemaking, it aims to improve the quality of life through an attractive design and a sustainable and quality built environment, where the historical and natural heritage and the urban landscape are enhanced. The plan considers that the distinctiveness of places has an influence on people's well-being as well as on health and social cohesion.

Likewise, high-quality design is also seen as a key element in enhancing the positive characteristics of Gateshead and Newcastle. For the creation of a successful place, it is important to fully understand the context from different points of view, namely socio-economic, cultural, natural and built. The distinctive character of the areas within Gateshead and Newcastle is given by its natural topography and landscape and the rural and urban ones around it. The views of these places and their enhancement can therefore be distinctive features for the quality of the place. Public art can also be of support in this direction, and if used in a sensitive and imaginative way, it can help improve the sense of place and/or improve the search for the road.

Specifically, the CS14 Policy deals with wellness and health to be maintained with initiatives that include creating a living environment favourable to older adults, creating an inclusive environment both natural and built, promoting healthy lifestyles, preventing impacts on residential services deriving from contamination of soil and water and air quality, providing adequate access to all health and social facilities and promoting accessibility for all the green spaces and spaces for sport and leisure.

The CS15 Place-Making Policy is aimed at creating environments that are attentive to the local character, safe, accessible, connected and inclusive and that respect the significant views and respond positively to the character of the River Tyne and its environment, the opportunities introduced by public art and local guidelines for design.

The CS16 Policy on Climate Change aims at sustainable development to be achieved with passive design, a good standard of the building fabric, flexibility from the outset to allow adaptability to alternative uses and resilience to the impacts expected from climate change and the risk of floods. The policy also envisages reducing $CO_2$ emissions and activating low-carbon energy solutions that comply with government guidelines. These guidelines will be followed by both the Gateshead and Newcastle Councils, and both adaptation and mitigation will have to be achieved by paying attention to the context, design and landscape, thus also improving their attractiveness.

The CS18 Policy, which concerns the quality and interconnection of green infrastructures and the natural environment, focuses on protecting and improving the connection and multifunctionality of green infrastructures; the protection and enhancement of natural corridors; green networks with particular landscape value and trees, woods and hedges; the extension and connections of the Green Infrastructure Network; the enhancement of open spaces and sports and leisure facilities; and the improvement of the axis to the River Tyne and tributaries.

Regarding the policies for the urban core, these focus on the quality of public spaces and paths and the protection and enhancement of the natural and built environment.

In particular, the UC12 Urban Design Policy underlines the importance of providing distinctive places in terms of quality of architecture and public spaces to be obtained through projects that enhance the positive characteristics of the context and the surrounding sites, improving primary and secondary paths and using durable and sustainable materials appropriate to the character of the area. The readability of the urban core must be clear and instinctive also thanks to suitable visual connections where there are high pedestrian flows, all following a coordinated and holistic approach to the site that respects its character and enhances or creates an appropriate urban form.

The UC13 Respecting and Managing Views Within, From and Into the Urban Core Policy calls for attention to prominent views such as those from, through and into the River Tyne Gorge; major roads; and major buildings and monuments of cultural interest.

The UC16 Public Realm Policy deals with the network of public spaces both regarding the improvement of existing ones and the creation of new ones in a logic of flexibility and adaptability to different uses, attractive design and use of high quality, durable and sustainable materials. This will improve the user experience, liveability and healthiness of the places, and thanks to the possibility of different and possibly temporary uses, it will increase the opportunities for social engagement (Figures 20.1 and 20.2).

The current presence of public art has created a strong sense of place, creating distinctiveness, strengthening the identity and attracting internal investments. The UC17 Public Art Policy intends to enhance local distinctiveness and the key sites where public art can be included. It can also be

*Figure 20.1* Newcastle-Gateshead, the project

(Source: HOK's archive)

*Figure 20.2* Newcastle-Gateshead, the project – evening view

(Source: HOK's archive)

integrated into the architectural fabric, the floor, street furniture, lighting or temporary installations and performances. Art can also help the legibility of the street by increasing the experience of the place and can be an opportunity for collaboration and creative work between people and maximization of the use of resources (Newcastle and Gateshead Council, 2015).

## The Quayside area

The regeneration of the waterfront of Gateshead and Newcastle that began in the mid-1990s is transforming its former industrial destination into a contemporary place for leisure and culture, attractive and creative, characterized by liveability and healthiness (https://www.c-mw.net; https://www.gateshead.gov.uk; https://www.meetnewcastlegateshead.com).

The redevelopment actions focus on two neighbourhoods, namely Stephenson Quarter and Quayside. The Stephenson district is located between Quayside and Central Station and is characterized by abandoned areas linked to the area's railway past, commercial and industrial buildings and vast free spaces in disuse. The regeneration began with the completion of the Life Science Centre biomedical research centre connected to a large science museum, the Metro Radio Arena for concerts and performances and the Central Square office building.

The other neighbourhood is Quayside, which is the focal point of the city and where the first bridge over the River Tyne was built (Carta, 2007).

In 1996, on the occasion of the Year of Visual Arts, attention had already begun to use public investments – together with those of private organizations – in the creation of new infrastructures such as the 'Sage Gateshead' (dedicated to music), "Dance City" (dedicated to dance), and "The Baltic" (contemporary art centre).

A further infrastructure that also became a symbol of the territorial dimension of the project is the Millennium Bridge over the River Tyne, a connection for pedestrians and cyclists to the Gateshead side.

Quayside has become one of the most vital and liveable parts of Newcastle thanks to the presence of spaces for the arts and socialization. Indeed, several public spaces have been created to connect residences and services, as well as a transport network with the use of eco-sustainable vehicles. Art, health, liveability and creativity are here understood as necessary ingredients for urban, economic and social regeneration: on one side with the creation of galleries and exhibition centres in the industrial buildings of the nearby cultural village Ouseborn and on the other side with the creation of a music school within the Sage, open to children and people of all ages.

The Quayside is crossed on both banks by the River Tyne along which there is a cycle/pedestrian path of great interest because of the views of the bridges, the architecture and the landscape, offering a notable perception of liveability and healthiness. Another element of liveability is given by Quayside Seaside, the urban beach that is set up in the summer near the Tyne Bridge, with sand, deck chairs and cabins. The equipment of this area for the summer includes a climbing wall, games for children and spaces for activities such as yoga and Pilates.

The views of the bridges include the Tyne Bridge and the Gateshead Millennium Bridge, which was designed by Wilkinson Eyre Architects and constitutes an element of attraction both for its architecture, the particularity that it is a bridge that tilts and the possibility of walking along it and biking it.

The architecture that can be observed along Quayside include the Sage Gateshead, the regional music centre of international standing created by the Foster + Partners studio and the Baltic Centre for Contemporary Art, created from the recovery and transformation of a former grain warehouse from the 1940s (Figures 20.3–20.5).

*Figure 20.3* Newcastle–Gateshead, detail of the project

(Source: HOK's archive)

*Figure 20.4* Newcastle–Gateshead, detail of the project – evening view

(Source: HOK's archive)

*Figure 20.5* Newcastle–Gateshead, detail of the project

(Source: HOK's archive)

A recent plan will further improve the liveability and attractiveness of this place. The Newcastle Gateshead Quays Development Plan was presented by Ask PATRIZIA – Ask Real Estate and Patrizia – and was approved by the Gateshead Council's planning committee in November 2020. The plan's projects include an arena, an exhibition and conference centre, a hotel and new public spaces. The general plan was entrusted to the HOK studio; AHR Architects is designing the hotel while the designed of the public spaces has been entrusted to the Planit-IE landscape studio.

The construction of the 58,500-square-metre complex was entrusted to Sir Robert McAlpine, while the ASM Global company will manage the campus for 12,500 people, with an area of 29,000 square metres and 7,000 square metres of flexible spaces for different types of events such as sports, conferences, exhibitions and festivals.

Public spaces are an important part of this project as these are places conceived with biodiversity and designed for leisure time. These spaces include elevated areas with views of the River Tyne, seating areas, bicycle supply throughout the area and charging points for electric vehicles within the new multi-storey car park.

The total cost of the project is about 260 million pounds, and the ultimate goal is to create infrastructure projects to improve economic growth and quality of life and increase jobs. Four thousand new jobs and an income of about 60 million pounds per year are expected. The Getting Building Fund, managed by the North East Local Enterprise Partnership, awarded 7 million pounds to create the link between north and south in the Baltic district and improve accessibility to the areas of the regeneration project. Two million pounds were allocated for the construction of the exhibition and conference centre, public spaces, the arena, two hotels and the multi-storey car park. The conference centre will be 2,800 square metres and will be able to accommodate a flexible number of people, the exhibition space will be 6,300 square metres and the arena for outdoor events will seat 12,500 people. The designed public spaces will cover an area of 2 kilometres, which includes roads, cycle paths, pedestrian crossings and green infrastructure. In this regard, the project includes seating areas and elevated areas with lookout points over the River Tyne, bicycle rentals and charging points for electric vehicles (https://www.c-mw.net/on-the-newcastlegateshead-waterfront-big-reveal; https://www.meetnewcastlegateshead.com/newcastlegatesheadquays).

## Conclusion

Since the mid-1990s, the regeneration of the waterfront of Gateshead and Newcastle has been transforming its former industrial destination into a contemporary place for leisure and culture, characterized by liveability and healthiness.

The new spaces of the Quayside neighbourhood are being designed following the principles of biodiversity and sustainability and planned to accommodate outdoor activities in order to solve the growing need for quality outdoor spaces also due to the COVID-19 emergency.

An important aspect of this project – which is about to be completed – is the flexibility in terms of both interventions to be carried out and the number of people to host. This aspect is useful for periods of health emergency and, in any case, to allow the realization of different types of outdoor activities: not only those related to physical activity and free time but also integrating congress, cultural and public art activities (https://www.ngquays.com/landscaping/).

In addition, about 144 trees will be planted to create green spaces where wildlife can live well and give life to the vegetation that will flank paths and spaces and places for community activities. Accessibility is provided 24 hours a day for everyone with walkways, elevators and escalators. The approach to the landscape was designed while paying attention to the history, geology, microclimate and structure of the site. The process of listening to people has interested the people who live, work, study and visit the surrounding places in order to better focus on the wishes and expectations of those who will use these spaces and encourage people to enjoy this part of the city.

The idea is to create a place to improve people's lives, have a positive impact on climate change and resources by 2050 and achieve zero-carbon buildings, thanks to the connection of Newcastle Gateshead Quays to an energy network that generates both energy through a sub-stream network of heat cables and high-voltage electrical cables using carbon.

The development plan also includes a social value plan to improve employment as well as education through internships provided to schools, colleges and universities.

Furthermore, the approach to sustainability is understood in its triple meaning and the goal of building sustainable environments has as its ultimate objective to improve the life and health of people and the planet.

# 21 Nice

## The area

Nice, a city in the French Riviera in the Provence–Alpes–Côte d'Azur Administrative Region, in the last years, has been involved in a wide operation of transformation which include new public spaces.

Accordingly, the Promenade du Paillon case study is of particular interest for the capacity of the area to be, at the same time, a place of reconnection between two important places of Nice and a place of entertainment and break of urban and landscape quality. Surveys, observations and questionnaires were carried out during a year and in different periods, both during the week and weekends during daytime and afternoon hours.

The first phase consists in the choice and definition of the study area. Two site inspections were carried out on the Promenade du Paillon public space and on the area close to the different accesses, and it was decided that the park, in its entirety, would constitute the study area. Indeed, it constitutes a place with characteristics of continuity which is diverse from the surrounding areas.

The boundaries which were identified concern Avenue de Verdun, which continues with Avenue Faurè on the right side and Boulevard Jean Jaurès on the left side, and from Promenade des Anglais on the southern part and to the Théâtre National on the northern one.

The second phase is characterized by the observation of the characteristics of the place. Surveys of the activities and relative modalities of the perceptions and elements which contribute in different ways to the liveability of the place were carried out.

Regarding the first survey, the typology of people (locals, visitors, professionals, etc.) and activities (passing by, working, taking a break, etc.) are observed. These activities are measured from a quantitative point of view, namely measuring how the specific activity is present in that place, how it influences the place's liveability and the frequency.

In order to illustrate the activities, the place is divided into five stretches, namely the stretch with the main access to the park from the Promenade Des Anglais, the stretch with the Arc sculpture and the first water mirror, the stretch with the second *water mirror*, the stretch with the natural lawn and

DOI: 10.4324/9781003098775-26

that with the synthetic lawn and the playground with wood games and the final stretch which ends at the National Theatre.

## People and activities

Surveys were carried out in the daily and afternoon hours both on weekdays and on the weekends. An average of results obtained in the different surveys is illustrated.

With respect to the activities of the first stretch, those mainly carried out include taking a break on the many wood benches and the passing by and the presence of access to the Promenade des Anglais. These activities are carried out mainly by residents and tourists of different ages (between 20 and 60 years) frequently.

Other activities which are present in minor quantity and frequency are the sport ones, namely jogging and observing the presence of the music kiosk, the fountain with the sculpture in white marble and the Theatre de Verdure. As to urban furniture, those which are present are wooden benches, the marble fountain and the open-air theatre.

The second stretch is characterized by the presence of a 20-metre-high iron sculpture titled *Arc 115°5* and a space with water games.

Here the activity is of passers-by on the path with the sculpture and the water games, with enjoyment by people using the little space with the water games. These games are constituted by water jets of various intensities which transform the pavement in a sort of swimming pool with low water or in a vapour cloud, making a dynamic and joyous space. The main and frequent activity is that of walking between water jets and the different transformations in the pool and vapour cloud.

Other less frequent activities are those of observing the urban scene of people who cross the water games and take pictures and videos. People who carry out the activities of crossing the water games are children and young people, while the other activities are carried out by people of different ages (between about 20 and 70 years), both locals and tourists.

In the third stretch, after that, a short pedestrian and car path have to be crossed; there is another pavement with water games of the same typology of the previous one but with major dimensions. This is bordered on the long sites by two pedestrian paths, of which one with wood benches and the other with a low construction of the City of Nice for tourist information and a shelter, in wood as well with benches. The activities which are present and very frequent include walking between the water games or in the different transformations in the pool and vapour cloud, observing this scene by people who cross these games, taking a break and taking pictures and videos. Also in this case, people who carry out the activities of crossing the water games are children and young people, while the other activities are carried out by people of different ages (between about 20 and 70 years), both locals and tourists.

The fourth stretch is characterized by the presence of a low lawn and trees which is followed by pavement and a synthetic lawn with wood games of great dimensions and quality design for children and a little pavement with water games with the same characteristics of the previous ones.

These two stretches are bordered by two paths: one with benches and the other with a low wood construction with toilette and, in the centre, an old marble-and-iron statue titled *Nice a Massena*.

The stretch of lawn is used in particular for picnicking and taking a break with medium frequency mainly by young tourists, while taking a break on the benches is carried out by people of all ages, both locals and tourists. The area with wooden games is used with high frequency by children and parents, both locals and tourists, for enjoyment and to take photos and videos of the children's activities. The lateral paths which border this stretch are used by locals, workers and tourists of all ages with high frequency. The little paving with water games is used only by children with a low-medium frequency.

Finally, the last stretch, which ends before the National Theatre, is characterized by a lawn with trees, benches, a little paving with water games and lateral paths on the two sides, one of which has a kiosk with multimedia support for recharging phones, tablets and the like.

In this stretch, activities which are present include taking a break on the lawn, using the pavement with the water jets, observing, walking and recharging mobile phones. Of these activities, the more present and frequent are taking a break and passing by on the lateral sides, both carried out by locals, tourists and workers.

The pace with which these activities are carried out is generally moderate and slow, offering a very agreeable perception of that place.

The second survey of single and mixed perceptions was carried out. Regarding single perceptions, all involved but taste. The visual perception is involved in a notable manner, in particular, in the first four stretches and concern: the view of the sea from the Promenade des Anglais access, that of the hill from the central stretch with the main pavement with water games, the trees which are present in many stretches, the sculptures and the various water effects offer a very agreeable perception.

The sound perception is present in a notable manner mainly in the stretch with the water paving of major dimension and in that with the wooden games, which are the most frequented.

For their characteristics, these create a perception of liveability and happiness expressed also by the high voice tones of surprise and joy who people use in these places. Smell perceptions, all agreeable, are stimulated by lawn, flowers, trees and all the plant species present in the park.

Finally, the tactile perception is stimulated by the different materials of paving, the lawn, the water games and the wooden games – all very agreeable.

The mixed perception which is mainly present is that of joy, in the stretches with both the water and wooden games and of serenity in the other.

Then the third survey is that of the elements which contribute more or less to the perception of liveability. The elements of the park which contribute to this perception include, for the natural elements, the lawn and the multiple plant and flower species which are present and the water in different shapes; the urban furniture; the paving with the water games; and the games that are realized in natural materials such as wood, including the turtle, the galleon and the whale (Figures 21.1–21.11).

*Figure 21.1* Nice, Promenade du Paillon, entrance by the Promenade des Anglais

*Figure 21.2* Nice, Promenade du Paillon, view with the *Arc de 115°5* sculpture

*Figure 21.3* Nice, Promenade du Paillon, view with the water games

*Figure 21.4* Nice, Promenade du Paillon, perspective of a path

*Figure 21.5* Nice, Promenade du Paillon, view with the water games

*Figure 21.6* Nice, Promenade du Paillon, view with the playground

*Figure 21.7* Nice, Promenade du Paillon, view with the playground

*Figure 21.8* Nice, Promenade du Paillon, view with the playground

*Figure 21.9* Nice, Promenade du Paillon, view with the Office of Tourism

*Figure 21.10* Nice, Promenade du Paillon, the picnic space

*Figure 21.11* Nice, Promenade du Paillon, natural path

## The places from peoples' points of view

Phase 3 concerned the questionnaire on-site of the users of the place, devoted to comprehending the elements which make that place liveable and healthiness from their points of view.

The on-site questionnaire was administered to about 110 people, locals and tourists, in English, French and Italian. Questions were asked mainly of people who were taking a break on the benches or on the lawn in the different stretches of the Promenade.

All the people interviewed answered the questions, although dedicating different times to answering.

The following questions, and their answers, are reported:

1. Does this place give you a feeling of happiness or sadness/liveability or discomfort/healthiness or unhealthiness?
2. What are the elements that give you the above sensations?
3. What are the main facilities that give quality to this place?
4. What are the activities that you do in this place, and how often?
5. What do you think about the presence of many or few people here? Do you think that it is capable to improve the pleasantness or unpleasantness of the place?

6. According to the place's healthiness, liveability and happiness, what could be done in order to improve this place?
7. What is a healthy/happy/liveable place that you remember in this city or elsewhere?
8. How might the weather condition influence the perception of healthy, liveability or happiness of this place?

To the question, "Does this place gives you a feeling of happiness or sadness/liveability or discomfort/healthiness or unhealthiness?" 100% of people – both locals and visitors – answered liveability and happiness. Regarding the second question, "What are the elements that give you the above sensations?" for the 80% of locals and tourists it was the landscape with trees, plants and so on, the water jets, the space in itself for different activities or for taking a break. The remaining 20% of people, mainly locals, answered that the element which makes it liveable is the clean air that is breathed while being at the centre of the city.

To the question, "What are the main facilities that give quality to this place?" 100% of people, both locals and tourists, answered the wooden playground which is scenic (above all mothers and children), the presence of many wooden benches (mainly mothers with children, tourists and older adults), toilettes and presence of pavement with the water games which allow for cooling off in warm periods as well.

To the question, "What are the activities that you do in this place, and how often?" answers were different but consistent with the possibility of activities which the place offers. One hundred per cent of those interviewed answered the amusement activity – meant as observing the urban and landscape scenes which the place presents, games and so on –which is carried out with frequency by almost all the interviewees. Of these interviewees, 20% of people added sport – namely running, which was done with a medium frequency, above all in the early morning hours; walking; and taking a break, the latter which was done at a high frequency by locals and a medium frequency by tourists.

To the question, "What do you think about the presence of many or few people here? Do you think that it is capable to improve the pleasantness or unpleasantness of the place?" 100% answered that the presence of a medium number of people increases the pleasantness of the place, giving an atmosphere of vitality to the park. Of these respondents, 50%, both locals and tourists, underlined that the pleasantness is given also by the presence of people of different ages.

Regarding the question, "According to the place's healthiness, liveability and happiness, what could be done in order to improve this place?" the answer given by the 100% of interviewees was nothing, denoting the positive perceptions the place generally gives.

To the question, "What is a healthy/happy/liveable place that you remember in this city or elsewhere?" answers concerned many places. Eighty per cent of French interviewees answered places such as gardens and public spaces in Nice and in the French Riviera, and the other 20% answered parks in Paris. The other tourists answered with different green spaces in their city, but all have underlined the uniqueness of the Promenade du Paillon. Finally, to the question, "How the weather condition might influence the perception of healthiness, liveability or happiness of this place?" answers were different. Of both locals and visitors, 80% answered that many areas are used all year, except during the rainy times, even though shelters with benches are present and allow the protection from the rain. The remaining 20% of people – mainly locals – answered that the pavements with the water games in the cold months are less used, although they are always pleasant to observe.

## The transformation of the area

The fourth phase is about the traditional analysis phase. The Promenade du Paillon is a public park of rectangular shape with an internal drawing – which is both articulated and regular – of 1,200 kilometres. This is situated in the Jean-Médecin neighbourhood named Carabacel, between the old and 19th-century Nice and rises before the Paillon River, which divided the city into two parts. After being recovered, the stretch where the river rises, a bus station and parking were realized. In the framework of the operative programme PNRQAD (Programme National de Requalification des Quartiers Anciens Dégradés), the local administration decided to realize a regeneration of the old Nice area with an improvement in housing, public spaces and mobility with interventions which regard Place Massena (created in 1849 and renovated in 2007), the Promenade du Paillon (inaugurated in 2013) and the Gare du Sud (inaugurated in 2019; Anru, 2007; Bonnet-Galzy, 2014; Mateo et al., 2012; www.nice.fr; www.nicecotedazur.org).

The Promenade, commissioned by Métropole Nice Côte d'Azur and designed by landscape architect Michel Pena with Zekton Hydraudesign and Atelier Coup d'Eclat, is an urban park financed by Métropole Nice Côte d'Azur (with the participation of the General Council of the Maritime Alpes for 3.5 million euro), for a cost of 40 million euro.

It presents 600 trees, 6,000 shrubberies and 50,000 little plants with essences from all over the world.

The park, beyond the landscape elements, presents elements of urban furniture and equipment, such as fountains with water games, nebulizers, kiosks and covered spaces for entertainment and a playground for children. Two lateral paths, of which one is wider and the other is narrower, border the long side of the park while the principal accesses on the short side make the park accessible from different points of the street.

## The liveability and healthiness factors

Phase 5 regards the identification of elements which contribute to the liveability and health of this place. This place constitutes a green area situated in the centre of Nice, between the old neighbourhood and the 19th-century one and between the sea and, in perspective, the hill. Elements which contribute to the perception of health regard both tangible and intangible factors. The drawing of the park conceived with many functions and accesses in different points gives the possibility to use it from many parts of the place.

These functions are devoted to people of all ages, from children to older adults, both locals and tourists.

The use of natural material for all, including the urban furniture, playgrounds, low structures with toilettes and information office, and the benches create a harmonic *unicum*, giving agreeable perceptions which are well joined to the landscape – with the multiple essences, plants and vegetation – of the park.

The water mirrors of the paving that people cross or look at create a very high perception of joy. There are spaces, breaks and paths which are well designed in a sequence which offers to all users a feeling or perception of serenity and joy, offering a high degree of liveability, happiness and health as well (Figure 21.12).

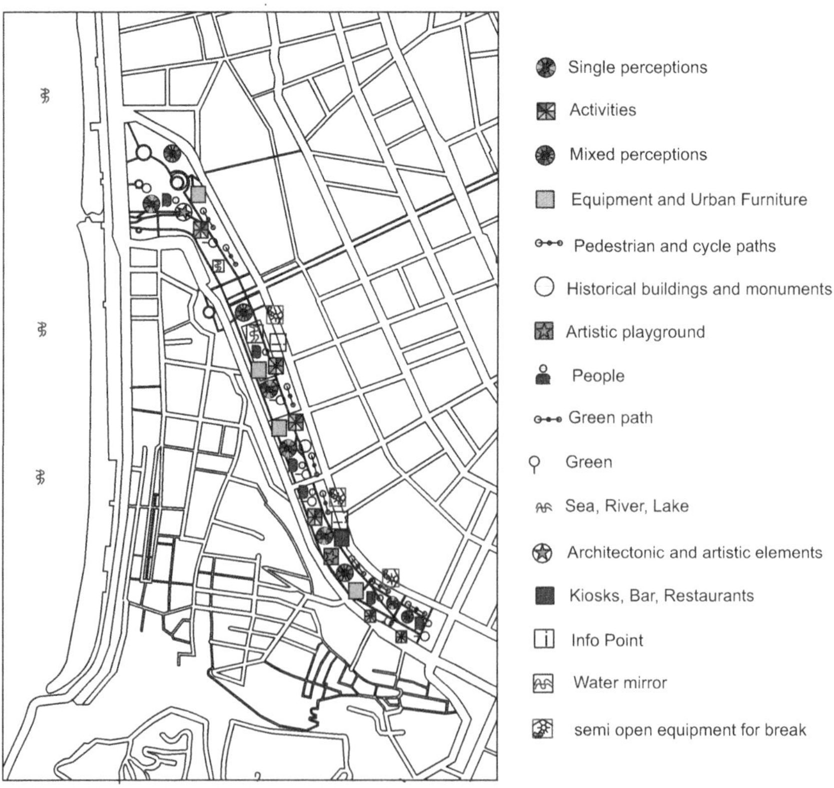


*Figure 21.12* Nice, Promenade du Paillon, map of urban health and liveability analysis

## Designing health and liveability

Checking the degree of health, liveability and happiness is carried out through the 25 principles of urban health and liveable design concerning the Charter. After checking the principles, the following could be improved: (7) to create suitable spaces for dogs and domestic animals; (9) to minimize or eliminate the noise generated by public transport; (10) to improve suitable cycle lanes; (16) to preserve both the place identity and the intangible characteristics of the site and its surroundings; (17) to both allow and promote different types of functions such as games, breaks, walking, and so on; (18) to facilitate gymnastic activities with the presence of small equipment or a designated space; (20) to encourage the presence of art in its different forms; (22) to promote participation, namely the feeling of being able to contribute to the life of that place increasing the sense of belonging; (24) to promote the educational function which a place has – for example clearly displayed information about the history of the place and the like or suitable ways to use it – increasing its intrinsic value; and (25) to facilitate the use of new technology to increase the knowledge of its intangible values and history, offering a more profound experience of the place.

The items of the questionnaire which follow derive from the principles which could be further satisfied.

The questionnaire was administered to about 130 people aged between 20 and 75 years, including locals and visitors from all over the world. The languages used were English and French, and people generally answered with pleasure.

1. What about adding a cycling- and a pedestrian-dedicated lane?
2. What do you think about inserting or enhancing artworks?
3. What do you think about adding some equipment for gymnastic exercises?
4. What about adding an area for dogs?
5. What about identifying suitable sites for participation events?
6. What do you think about minimizing the sound of vehicles?
7. What about adding information about the area and using new technologies?
8. What about creating or enhancing spaces for educational function?

With respect to the first question, "What about adding a cycling- and a pedestrian-dedicated lane?" 80% of people – both locals and visitors of all ages – answered positively. Of these, local people added that the cycle lane could be enhanced in the external part of the Promenade du Paillon, beyond the theatre and the museum as well. The remaining 20% of locals and visitors answered that they agreed, adding that the lines should be realized in natural material in continuity with the landscape of the Promenade.

With respect to the question, “What do you think about inserting or enhancing artworks?” 70% of locals and visitors of all ages answered that they agreed both with creating new artworks and enhancing some statues which are already present in the Promenade, such as the Arc situated at the beginning of the park. The remaining 30% of locals and visitors of all ages answered that they agreed and that the artworks could be realized in different ways, including the elements of nature, such as flowers, wood and so on.

With respect to the question, “What do you think about adding some equipment for gymnastic exercises?” 80% of interviewees – both locals and visitors of all ages – answered that they agreed with this idea. The remaining 20% answered that they agreed but that the equipment should be light, in line with the landscape of the Promenade and for all.

Regarding the question, “What about adding an area for dogs?” 90% of both locals and visitors of all ages answered that they agreed with adding an off-leash area. Of these interviewees, locals added that this could be equipped with games for dogs and benches for their owners. The remaining 10% of people generally agreed with the idea, adding the importance of having a site for dogs also for rainy or very warm periods.

Regarding the question, “What about identifying suitable sites for participation events?” 80% of locals and visitors agreed with this idea. Of these respondents, locals added that participation events could be organized in the parts of the Promenade closer to the accesses. The remaining 20% of locals and visitors generally agreed with this idea, although they thought that participation could be organized through blogs or dedicated web portals (Figure 21.13).

With respect to the question, “What do you think about minimizing the sound of vehicles?” 100% of locals and visitors answered that they agreed with reducing it in the parts of the Promenade where the means of transport are closer or streets cross the area. Of these respondents, 50% of locals added that they would prefer the pedestrianization of those parts to allow a better continuity to the park.

Regarding the question, “What about adding information about the area and using new technologies?” 80% of people – both locals and visitors – answered positively, adding that the information could include a map of the Promenade with details about the plants and the other elements of the place, activities and equipment as well as transport timetables, weather forecasts and so on. Younger people added that a small area dedicated to Instagram photos could be inserted. The other 20% generally agreed with this idea and would prefer the use of new technologies.

With respect to the question, “What about creating or enhancing spaces for educational function?” 80% of people – both locals and visitors of all ages – answered that they agreed, adding that they would be interested in the history of the place, the natural environment of the Promenade and knowledge of how the place contributes to sustainability. The other 20%

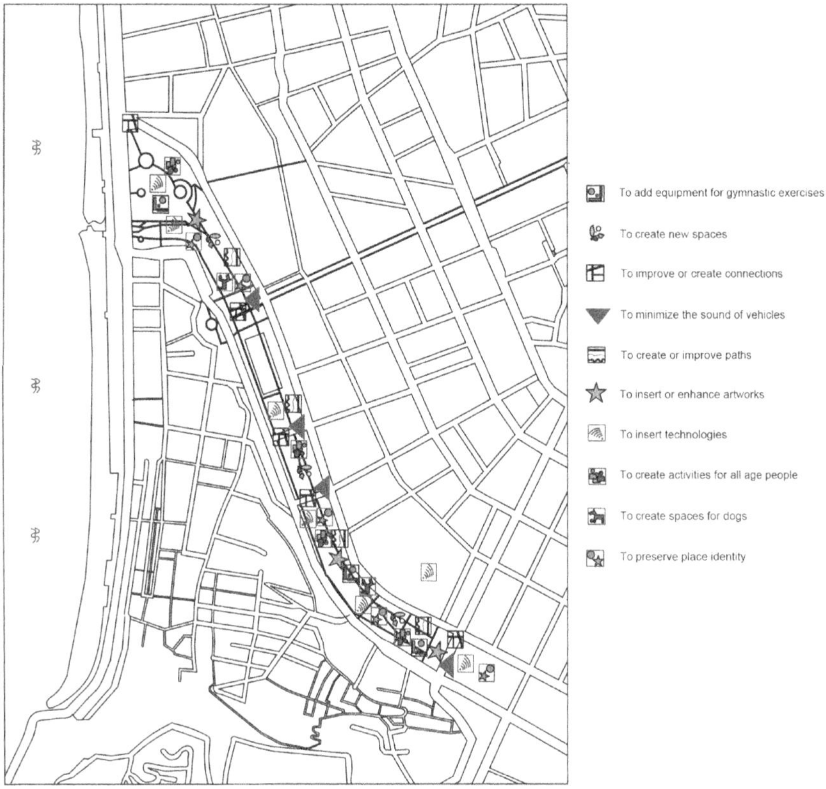


*Figure 21.13* Nice, Promenade du Paillon, map of urban health and liveability design

answered that they generally agreed, adding that the educational functions could be supported by the new technologies.

Researching reviews through websites, the first data which emerge were the high presence of the Promenade on social networks. Regarding Instagram, the hashtags used include #promenadedupaillon, with 8,984 posts, and #promenadedupaillonnice, with 193 posts. With respect to Facebook, there are two un-official Promenade du Paillon pages, one with 1,104 followers and another with 449 followers. On Twitter, there is no official page, although the #PromenadeDuPaillon hashtag is present within many tweets. Finally, among booking websites which report a major number of reviews, TripAdvisor, with 1,012 reviews, focused on the Promenade. The prevailing reviews are positive from all the points of view and, in particular, the beauty of the water mirrors is highlighted; in the comments there is evidence not only of the aesthetical aspects but also those concerning the safeness and clearness of the whole Promenade. Another interesting point is that those reviews are provided during the whole year, also evidencing that during the Christmas period, there are particular lighting and water games suitable to that period of festivity. Many people would like it if this park were built in their city.

From the information gathered in the previous phases of the project, the following project interventions were designed (Figure 21.13).

The first project intervention is creating new spaces or enhancing existing ones. The Promenade du Paillon is a green axis with the possibility of many activities. To further increase the healthiness, liveability and happiness of the place, its extension could be improved on the north side beyond the Museum of Modern and Contemporary Art, creating a wider green area for observation spaces, picnic sites, an open-air theatre and so on. Furthermore, a participation area could be identified to increase the feeling of belonging at the place but also temporary and new uses of the Promenade. These places could be inserted in the current Promenade, or more spaces could be found, extending the area toward the Acropolis.

The second project intervention is the enhancement or creation of connections. The continuity of the Promenade is currently interrupted in some parts, including at Place Massena, Allée Résistance et Déportation and Trav. É Flandres-Dunkerque, where the last two are crossed by cars. To improve the continuity of the promenade, connections between these stretches could be reinforced with their pedestrianization. Connections could also be improved in Avenue de Verdun and Trav. de la Bourgada, namely, respectively, at the beginning and the end of the Promenade, which are used less than the entrances to the central part.

The third project intervention is creating a cycling-dedicated lane. The realization of a bike lane – in the framework of a pedestrianization of the car lines which interrupt the continuity of the Promenade – could constitute a healthy path through the landscape. The possibility of continuing it beyond the end of the park until the Acropolis could further improve the power of the Promenade as a green axis. To create a suitable and safe path for cyclists, it should be separated from the pedestrian ones.

The fourth project intervention is differentiating activities for different age groups. Playgrounds for children and water mirrors are already present in the Promenade. To introduce further games for people of all ages – including older people – and abilities could improve the health, liveability and happiness of the place.

The fifth project intervention is adding an area for dogs. The creation of a dedicated off-leash area could allow dogs and their owners to freely go around the site. The dog area could have equipment and amenities in natural material and be in line with the landscape of the place to improve the liveability and happiness of the Promenade.

The sixth project intervention is preserving or enhancing the identity of places and intangible cultural resources. In the Promenade, there are fountains and sculptures, including the *Arc de 115°5*, the *Statue de David*, the *Fontaine Des Trois Grâces* and the *Statue du Maréchal Andre Masséna*. These – including the place's identity and the cultural resources nearby – could be further enhanced in the Promenade with dedicated spaces around them and/or information through QR codes.

The seventh project intervention is minimizing the sound of vehicles. The sound of vehicles mainly come from the parts of the area interrupted by the crossing of cars and from the parts of the Promenade closer to access points. To minimize the sound and make this place healthier and even more liveable, the pedestrianization of some stretches, including the Allée Résistance et Déportation and Trav. É Flandres-Dunkerque, could be carried out. Alternatively, a green barrier can be realized.

The eighth project intervention is adding equipment for gymnastic exercises. The equipment should be light, made of natural material and be in line with the place identity related both to the natural and the built environment. At the same time, these should be for people of all ages and abilities and could be related to specific paths along the Promenade.

The ninth project intervention is inserting artworks. Sculptures, creative gardens and light art, both permanent and temporary, could be added to the current artworks, further improving the liveability of the place. The artworks should be in line with the identity of the place and could be connected with the exhibitions of the other museums in Nice.

The tenth project intervention is adding wireless information points. The information to add could concern the artworks, the history of the Promenade and its surroundings, the equipment, the playground, the activities and so on. These could be contained in QR codes or dedicated apps contributing to the liveability of the place and its attractiveness.

## Conclusion

The operations of transforming the Promenade du Paillon, integrated with the regeneration projects which are interesting Nice in the last 20 years, are improving its attractivity and liveability.

The Promenade is, in particular, conceived with many functions and access at different points, offering the possibility of using it from many parts of the place. These functions are devoted to people of all ages, including locals and tourists.

The use of natural material for all the furniture, playground and the low structures for toilettes and information office and benches creates a harmonious unicum, giving agreeable perceptions which are well connected with the landscape and its multiple essences, plants and vegetation.

The water mirrors that people cross or look at create a very high perception of joy. There are spaces, breaks and paths which are well designed in a sequence which gives to all users a perception of serenity and joy, offering a high degree of liveability.

In the last years, parks have begun including more kinds of spaces dedicated to specific activities to improve their healthiness, liveability and happiness to invite people to take part in open-air motor activities and gymnastics, breathe clean air and, more generally, take a break.

Accordingly, the project interventions which were identified include creating new spaces or enhancing existing ones, enhancing or creating connections, creating a cycling-dedicated lane, differentiating activities for different age groups, adding an area for dogs, preserving or enhancing the identity of places and intangible cultural resources, minimizing the sound of vehicles, adding equipment for gymnastic exercises, inserting artworks, and adding wireless information points.

These project interventions should be meant in a logic of integration and flexibility, namely connected among them and – where possible – resilient to changes. Furthermore, to improve the continuity of the Promenade, its extension to the Acropolis would empower its peculiarity as a green axis. In this way, also due to its central position and the proximity to the sea, it would become further attractive to all.

# 22 Vancouver

## The areas

Vancouver is a city in the British Columbia Province of Canada. It has a particular landscape given by mountains on one side and sea on the other, all surrounded by a good balance between natural and built environment.

This case study is constituted by the new Coal Harbour and False Creek areas, which contain many elements of interest including the presence of public spaces, the sea, mixed uses and many activities for free time. The outputs of the Coal Harbour and False Creek case studies are reported in the next paragraphs, starting from the first one.

The on-site visits – according to the first phase of the Healthy Pl@ce Design method – confirm the Coal Harbour area in its whole as the place of analysis. In particular, Coal Harbour starts at Canada Place to the east and ends at the beginning of Stanley Park to the west; it is bordered to the north by Burrard Inlet and to the south by West Georgia Street.

## Coal Harbour. People and activities

Coal Harbour is a pedestrian area. Observations were carried out both during the week and weekends; the data which were collected are synthetized in the following.

The activities (phase 2 of the method) include walking, cycling, renting and riding bikes, resting, watching the panorama, eating and taking photos and selfies; making a tourist flight tour with the seaplane, attending conferences, workshop and exhibitions; and participating in personalized boat tours; all done with high frequency by the people who use this place.

Each activity has its own space; for example there is a lane for pedestrians, one for cyclists and skateboarders in both directions, suitable grass with benches to take a break and so on. Furthermore, there are many places to eat and drink, a convention centre, the Harbour Flight Seaplane Terminal and both boat and bike rentals.

DOI: 10.4324/9781003098775-27

Furthermore, there are artistic sculptures of different types – including *The Drop*, a steel 20-metre-high sculpture, a sort of homage to the power of nature which represents a landmark of the place – the aluminium LightShed – and a little garden with images of the Coal Harbour and Vancouver history.

The places are of very good quality and materials and the spaces devoted to all the activities are suitable, giving the general perception of a healthy, liveable and happy place. And, not last, this area gives a perception of safety.

Pleasant perceptions are, in particular, visual, smell and touch due to the beautiful views of both the harbour and the mountains, the presence of grass with trees and flowers and the well-maintained and comfortable floor. The only perception which is in some moments annoying is constituted by the noise of the helicopter engines which are used for the touristic tours.

People are constituted by professionals of many kinds – here there are many offices, hotels, a harbour and a huge convention center – inhabitants and visitors from all over the world. The different functions which are present in the area are used with high frequency by people of different ages, proveniences and professions.

## The places from peoples' points of view

The answers (phase 3) to the on-site questionnaire, administered in English language and given to respondents – about 130 users of the place from all over the world aged between 20 and 75 years – were quite similar with respect to the perception of happiness, liveability and healthiness given by this place.

The following questions, and their answers, are reported.

1. Does this place give you a feeling of happiness or sadness/liveability or discomfort/healthiness or unhealthiness?
2. What are the elements that give you the above sensations?
3. What are the main facilities that give quality to this place?
4. What are the activities that you act in this place, and how often?
5. What do you think about the presence of many or few people here? Do you think that it is capable to improve the pleasantness or unpleasantness of the place?
6. According to the place's healthiness, liveability and happiness, what could be done in order to improve this place?
7. What is a healthy/happy/liveable place that you remember in this city or elsewhere?
8. How might the weather conditions influence the perception of the healthiness, liveability or happiness of this place?

To the question, "Does" this place gives you a feeling of happiness or sadness/liveability or discomfort/health or unhealthy?" 100% of the respondents affirmed that they feel sensations of joy and strong liveability there, and they think that the place is built to be healthy.

Regarding the question, "What are the elements that give you the above sensations?" 50% of locals and visitors of all ages answered the landscape, the good-quality pavement and its wide surface and the general atmosphere; the other 50% answered the possibility of doing many activities or taking a break in a very beautiful environmental context.

With respect to the question, "What are the activities that you act in this place, and how often?" 60% of locals – aged between 35 and 75 years – answered walking, sitting on the benches, watching the panorama and taking a break; the other 40% in the same age range answered walking with the dog, eating or drinking something, cycling, running and taking the own boat for sailing. The locals aged between 20 and 35 answered walking, taking a break, cycling, running, skating, going roller, skating and meeting friends. Of visitors of all ages, 100% answered walking, cycling, watching the panorama, attending a conference, making a touristic tour and eating or drinking something.

With respect to the question, "What are the main facilities that give quality to this place?" about 100% of respondents, both locals and tourists, answered the good mix between the beautiful environment, art, quality street furniture and the restaurants and bars. The locals added the sport furniture and the good quality of the pavements.

To the question, "What is a healthy/happy/liveable place that you remember in this city or elsewhere?" 30% of people – mainly tourists – answered that they were surprised by the beauty of the panorama and the comfort of the streets; 30% of people – both locals and tourists – answered "Stanley Park", and the other 40% answered both "East False Creek" and Granville, all in Vancouver. Of this, 20% of inhabitants – mainly young adults – added other Vancouver places, including "False Creek South" and Yaletown.

To the question, "What do you think about the presence of many or few people here? Do you think that it is capable to improve the pleasantness or unpleasantness of the place?" 100% of people answered that the presence of many people gives pleasantness to the place, because the place is wide and people can stay comfortably in different parts of Coal Harbour.

With respect to the question, "According to the place's healthiness, liveability and happiness, what could be done in order to improve this place?" 100% of people answered nothing because the place is beautiful, liveable and full of activities to do. Thirty per cent of locals – mainly adults and young adults - added creating an off-leash area for dogs, and another 30% added reducing the noise of the seaplanes (Figures 22.1–22.8).

*Figure 22.1* Vancouver, Coal Harbour, perspective

*Figure 22.2* Vancouver, Coal Harbour, view with *The Drop* sculpture

*Figure 22.3* Vancouver, Coal Harbour, perspective with the path

*Figure 22.4* Vancouver, Coal Harbour, view with the *LightShed* sculpture

*Figure 22.5* Vancouver, Coal Harbour, detail of the path

*Figure 22.6* Vancouver, Coal Harbour, detail with the benches

*Figure 22.7* Vancouver, Coal Harbour, detail with the fountain

*Figure 22.8* Vancouver, Coal Harbour, the path to Stanley Park

## The transformation of the area

Phase 4, the traditional analysis, has resulted as follows. Coal Harbour and False Creek are new areas of urban, socio-economic and environmental regeneration in which the City of Vancouver (2021a, b, c) is investing much energy to improve both liveability and competitiveness. During the 1970s and 1980s (Mathewson and Karaga, 2018), Vancouver was interested in a migration of the middle classes from the inner city to the suburbs. The result was economic stagnation which finished with the urban regeneration investments for the 1986 World Exposition on Transportation and Communication. The development revitalized the downtown into a mixed-use area that helped obtain a wider regeneration of the rest of the city (Mathewson and Karaga, 2018).

Many new policies were utilized for the new vision of the historical centre contained in the "Living First Strategy". This is focused on the improvement of the quality of life, walkability, public transport, local services and quality public spaces. In addition, other tree factors have been introduced in order to regulate the planning of Vancouver, namely "the promotion of tall buildings in key city centre locations, the preservation of important views to water bodies and mountain's encircling the city, and the promotion of the development of buildings with setbacks stepping down to the harbour and False Creek". These policies – all related to both the height limitation of tall buildings and the protection of the view corridors – include the downtown official development plan (1975), the special review processes (2002) of the higher building advisory panel, the downtown capacity and view corridor study (2008), the historic area height review (2011) and the higher building review, the View Protection Guidelines (2011) and the General Policy for Higher Buildings (2014).

In Vancouver, the City Council is independent of the provincial legislature. For this reason, it can decide to realize a specific scheme concerning additional height or density although against the planning rules, if the benefits capacity is demonstrated. The accepted new design of view will result in a rezoning of the interested area and corridors.

This system brings benefits to different subjects, such as the city, the local community and the developers of the project. Each change is carried out with the idea to conserve and protect heritage and natural environment.

The new area of Coal Harbour, in particular – situated between Stanley Park and the Canada Place Pier – has been regenerated since 1990. Coal Harbour was an industrial area with shipyards, seaplanes and the railroads which connect the industrial areas with Canada. The name comes from the 1862 discovery of coal in the harbour. Before that, the area was an upper-class residential district known as Blueblood Alley for the large mansions which were built there.

These industrial activities ended after the Second World War; only in 1985, with the Canada Place Convention Centre did Coal Harbour started a new period thanks to its attractiveness. In 1990, a new regeneration plan was carried out with the aim to reach about 1 hectare of open-air spaces for every 1,000 people. This objective has been carried out in 15 years – beginning in

1995 – and the site is now constituted by streets and residences and many green areas and public spaces. The area is limited by Stanley Park from a side and Canada Place from the other, and the waterfront constitutes the lateral border (Mckenzie and Hutton, 2015; Punter; 2003; https://vancouver.ca).

## The liveability and healthiness factors

The main factors which are reported in phase 5 of the method – the map of urban health, liveability and happiness – are constituted by the suitable balance among a beautiful natural environment – visible from the whole area, the variety of activities which can be done – both free and for a fee, the good quality of street furniture – well maintained and managed, the presence of art – above all sculptures – which give a good and harmonic mix of new and old and safety perception – also demonstrated by the fact that people of different age and motorial abilities use this area during all the day hours.

People use this area because can find here many activities to do in a beautiful both natural and built environment. Furthermore, this is a place for all ages; local people come here for relaxing, walking, cycling, running, watching the panorama, while visitors come here for these activities and for the touristic ones which are many and varied. This place is perceived as healthy by the all the interviewed, and this is achieved through a long regeneration process which has interested Vancouver, aimed at improving its liveability as a first factor, as in the case of the "Living First Strategy".

Furthermore, each activity is present in the most comfortable manner, including large, double-sense, separated and well-equipped lanes to walk and do other activities; many benches in front of the sea; good-quality pavement; and much information on the place (Figure 22.9).

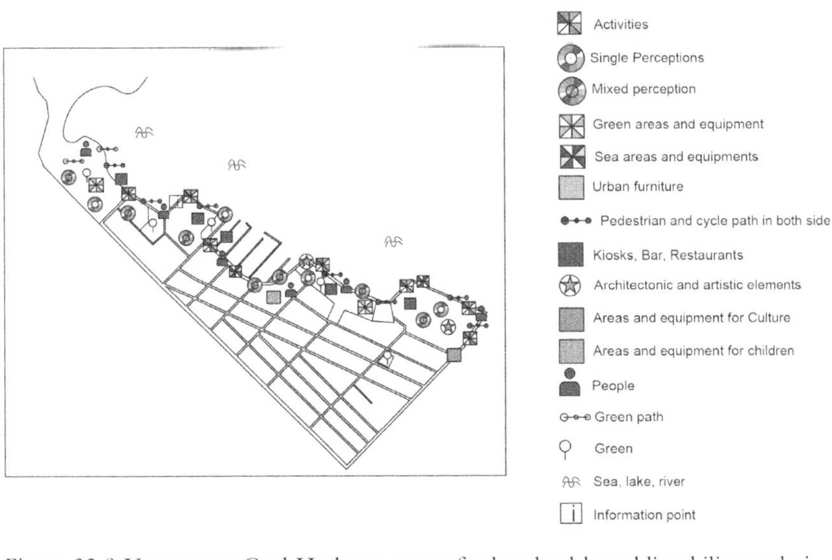


*Figure 22.9* Vancouver, Coal Harbour, map of urban health and liveability analysis

## Designing health and liveability

By checking the consistency between the results of different surveys and the Charter of Urban Health and Liveable Design, Coal Harbour area is in line with almost all.

The principles that could be improved include (2) to encourage the use of the place by people of different age groups, from children to older adults; (7) to create suitable spaces for dogs and domestic animals; (9) to minimize or eliminate the noise generated by public transport; (15) to have the possibility of using the space in different weather conditions and seasons, contributing at the same time to its good state of maintenance; (18) to facilitate gymnastic activities – also slow – with the presence of small equipment or a designated space; and (20) to encourage the presence of art in its different forms.

Once the principles are identified, the questions which follow derive from the principles and the results of the previous phases.

The questionnaire was administered to about 150 people, both local and visitors from all over the world, aged between 25 and 75 years. The languages used were English, and people were willing to give answers to the questions.

The following questions, and their answers, are reported.

1. Do you use this area during all periods of the year? What do you think to create urban public light structures to use this place also in very cold or warm periods?
2. What about inserting old or contemporary art products in their different forms to contribute to the sense of healthiness, liveability and happiness of this place?
3. What do you think about widening the game areas by inserting more playgrounds and places for games for people with disabilities and older adults?
4. What do you think to insert more equipment for open-air gymnastic activities also for people with disabilities and older people?
5. What about minimizing the sound of vehicles and seaplanes in the parts where this is more evident?
6. What about creating an area devoted to dogs?

To the first question "Do you use this area during all periods of the year? What do you think to create urban public light structures to use this place also in very cold or warm periods?", ninety per cent of people independently by nationality and age answered positively, although some of them added that during the coldest periods with the snow the structures should be stronger than in the warmer periods. The remaining 10% answered – above all locals – that for these periods there are many cafes and restaurants that could be used during uncomfortable weather conditions. Of visitors, 10% answered that they prefer not to go to a park or public space if the weather conditions are too unpleasant.

As regards the second question, "What about inserting old or contemporary art products in their different forms to contribute to the sense of healthiness, liveability and happiness of this place?", 100% of locals and visitors agreed with this idea, answering that in Coal Harbour, there are already sculptures of great dimension, but some others along the path could be agreeable.

To the third question, "What do you think about widening the game areas by inserting more playgrounds and places for games for people with disabilities and older adults?", ninety per cent of local people and visitors of all age answered that they agreed with the widening of the playground and game area. Many of them added that they would like wider areas for not only older people and people with disabilities but also for children. The other 10% of local people answered that they agreed with this idea but that for them, it is easier to widen the existing playground than create a new one. The remaining visitors did not have a specific idea about it; they enjoyed Coal Harbour and answered that there are already many activities to do there.

With respect to the fourth question, "What do you think to insert more equipment for open-air gymnastic activities also for people with disabilities and older people?" 80% of local people of all age answered that they agreed with this idea; many of them added that this equipment should be usable also in snowy or rainy periods. The other 20% of locals generally agreed, adding that they prefer only to walk or go by bike. One hundred per cent of visitors also generally agreed with this idea; 10% of them added that they would prefer equipment made of natural materials and suitable for warm and cold periods; another 10% added that they agreed, although they did not practice gymnastic activities.

As regards the fifth question, "What about minimizing the sound of vehicles and seaplanes in the parts where this is more evident?" 70% of both locals and visitors, answered that they are annoying only by the sound of the seaplane departure. The other 30% answered that the sound of the seaplane is not particularly annoying; they are disturbed in some parts of the area by the sound of vehicles.

Finally, to the question, "What about creating an area devoted to dogs?" 100% of both locals and visitors were positive about this idea, answering that there are few off-leash areas for dogs. Thirty per cent of locals – mainly adults and young adults - added that they would like some dog game structures as well.

Checking social networks and websites with suggestions, the Instagram hashtags used include #coalharbour, with 137,038 posts; #coalharbourvancouver, with 2,491 posts; and #coalharbourseawall, with 1,617 posts. On Facebook, there is "Coal Harbour Marina", "Coal Harbour Resident Association" and other pages that show images and information on this area, although there is no official page for the whole area.

Regarding the information collected by the internet websites, on TripAdvisor, there are about 88,490 positive comments about this place and its attractions. On Booking.com, there are hundreds of positive comments both on the area itself and the activity options there.

From the information gathered in the previous phases of the project, the following project interventions were designed.

The first project intervention is the creation of new spaces. The new spaces should concern playgrounds and game areas to be used by people of all ages, including children, older adults and people with disabilities. Coal Harbour is a place with many activities, and the creation of new areas and/or the widening of playground spaces for games and free-time activities devoted to children, older adults and people with disabilities could furtherly improve the perception of healthiness and liveability of this part of Vancouver (Figure 22.10).

The second project intervention is the creation of spaces for different climatic conditions. The spaces could be created with the use of light flexible structures that can be moved when these are not necessary. These structures could be useful both in very warm periods or when there it is snow or again when it is very cold and could be used for breaks or activities of different kinds.

The third project intervention is the insertion or enhancement of artistic elements. Sculptures of great dimensions are already present in Coal Harbour including the *LightShed*, *The Drop* and the *Digital Orca* and constitute an important element for the urban health of this place. Some more artistic elements which can be suitably seen and perceived by locals and visitors could be added to further improve the Coal Harbour's liveability. Furthermore, regarding those already existing, the public space around them could be enhanced in order to better admire the art elements.

The fourth project intervention is to differentiate activities for different age groups. Activities for locals and visitors in Coal Harbour are of many kinds. A further improvement could be the differentiation of activities for different age groups and, in particular, children, older adults and people with disabilities. This project intervention is connected to the creation of new spaces; both new spaces and related activities should be flexible, connected among them and attractive.

The fifth project intervention is to insert new equipment. These should include equipment for both physical and gymnastic activities for people of all

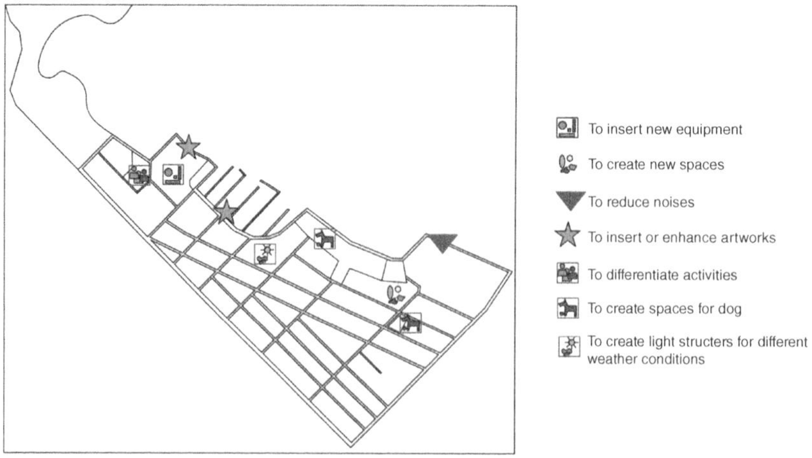


*Figure 22.10* Vancouver, Coal Harbour, map of urban health and liveability design

ages and people with disabilities. The equipment should be inserted where there are easy accesses, above all for people with reduced or slow movements.

The sixth project intervention is to create suitable spaces for dogs and pets. Dogs and pets should have suitable off-leash and devoted areas with games and space suitably inserted in the nature. The spaces should have close trails for walking, running or other activities for the dog and pet owners, improving the health, liveability and happiness of the place for both dogs and people.

The seventh project intervention is the reduction of vehicle noise. The main part of Coal Harbour is close to sea; only the internal part is crossed by some driveways. The noise of vehicles perceived from there and of the sea-planes perceived from the waterfront could be reduced by creating suitable green barriers.

## False Creek. People and activities

This area contains many elements of interest including the presence of public spaces, a playground, monuments, the waterfront, panoramas and many mixed-use and free-time activities.

Regarding the first phase of the Healthy Pl@ce Design method, with the on-site visits, the borders of the area were decided. False Creek is the area delimited by the Science World British Columbia and English Bay. After the on-site visits, it was decided that the study area would the body of water between Burrard Street bridge and Quebec Street. The waterfront area concerning the False Creek basin is mostly a pedestrian area. Observations were carried out both during the week and weekend; the data which were collected are synthetized in the following.

The activities (phase 2 of the method) are many and different due to the extension of the place and the existence of many functions.

The area – to both better collect the data and illustrate the results of the surveys – is described in three parts, namely North, South and Olympic Village.

The North part is between Burrard Street and Cambie Street; the South walking route stands along the seawall from the Cambie Street Bridge and the Granville Street Bridge. The Olympic Village is between Cambie Street and Quebec Street.

In the North area, the activities include walking, cycling, running, resting, watching the panorama, eating, parking and taking yachts or ferries, and taking photos and selfies, all done with a high frequency and at a general moderate pace by the people – locals, professionals and visitors -who use this place (Figures 22.11–22.19).

In the South area, the activities include walking, cycling, running, resting, eating, going to Granville Island, watching the panorama, taking photos and taking the shuttle ferries. These activities are carried out with a high frequency by locals, professionals and visitors who use this place at a moderate pace. Other activities include going to the gym and walking babies in strollers, which are carried out with a medium-high frequency mainly by locals at slow-moderate pace.

In the Olympic Village, the activities include visiting the Science World Museum, visiting the Village and participating in sports, walking, cycling,

*Figure 22.11* Vancouver, False Creek, detail of the path with the Science Centre

*Figure 22.12* Vancouver, False Creek, detail of the path with the beach and the benches

*Figure 22.13* Vancouver, False Creek, detail with the beach and the lawn

*Figure 22.14* Vancouver, False Creek, perspective

*Figure 22.15* Vancouver, False Creek, detail with a playground under a bridge

*Figure 22.16* Vancouver, False Creek, Granville Island

*Figure 22.17* Vancouver, False Creek, detail with chairs and a ferry boat stop

*Figure 22.18* Vancouver, False Creek, detail of the path

*Figure 22.19* Vancouver, False Creek, detail with engaging sculptures

running, canoeing, kayaking, resting and playing in the playground. These activities are carried out with a medium-high frequency and at a moderate pace by both locals and visitors.

Single pleasant perceptions are, in particular, visual, touch and smell, due to the panorama and views of both the basin, the presence of sand and grass with trees and flowers and the well-maintained and comfortable floor. In Granville Island the perceptions include the taste as well smell due to the various particular food products which are in the market.

The mixed perceptions are related to the fact that the places are of very good quality design and materials and the spaces devoted to all the activities are suitable. Then, the mixed perceptions include healthiness, liveability and happiness. In many parts of False Creek, there are sculptures of different types, which improve the perception of happiness often connected to the art. Furthermore, this area gives a perception of safety.

People who use this place – due to the presence of many activities and facilities – are mainly locals, residents, professionals and visitors of all ages. Furthermore, people with disabilities have here comfortable or dedicated paths.

## The places from peoples' points of view

Answers (phase 3 of the method) from people – about 100 locals, residents, visitors and professionals, with the questionnaire administered in English – were quite similar with respect to the perception of liveability, healthiness and happiness given by this place.

The following questions and their answers are reported.

1. Does this place give you a feeling of happiness or sadness/liveability or discomfort/healthiness or unhealthiness?
2. What are the elements that give you the above sensations?
3. What are the main facilities that give quality to this place?
4. What are the activities that you act in this place, and how often?
5. What do you think about the presence of many or few people here? Do you think that it is capable to improve the pleasantness or unpleasantness of the place?
6. According to the place's healthiness, liveability and happiness, what could be done in order to improve this place?
7. What is a healthy/happy/liveable place that you remember in this city or elsewhere?
8. How might the weather conditions influence the perception of healthiness, liveability or happiness of this place?

To the first question, "Does this place gives you a feeling of happiness or sadness/liveability or discomfort/health or unhealthy?" about 100% of interviewees answered that the place gives positive feelings such as happiness, liveability and healthiness.

To the question, "What are the elements that give you the above sensations?" answers covered different ideas of the place due to the many functions which are present here. Sixty per cent of both locals and tourists of all age answered that they are surprised by the beauty of the panorama and the comfort of the streets; 30% of people – mainly locals – answered that they like the possibility of the many kinds of things there is to do and use the area at different hours of the day and in different weather conditions.

To the question, "What are the main facilities that give quality to this place?" Of both locals and visitors, 70% answered that there are lanes dedicated for walking, running, cycling and skating, often in both directions. Of this group, 30% of people – mainly adults - added that this represents an important factor for safety and comfort. The other 30% answered different facilities, including mainly bars and restaurants and then benches, playgrounds, open-air gym furniture and a museum.

To the fourth question, "What are the activities that you do in this place, and how often?" 65% – both locals and visitors, mainly adults and young adults – answered that they walk, take a break, go by bicycle or run. The remaining 35% perform different activities including – for locals – bringing children to the playground or a parent in the wheelchair, going to the Granville market to buy food or objects, and going to the museum, taking their boat for a ride, taking the shuttle boat, going to the restaurant or pub and so on in the area, for both locals and visitors.

To the question, "What do you think about the presence of many or few people here?" 100% of both locals and visitors answered that the presence of many people gives the place a good atmosphere of liveability and pleasantness.

To the question, "According to the current place's healthiness, liveability and happiness what could be done in order to improve this place?" 65% of locals answered that to improve the liveability of the place, some parts have to be completed, and the remaining 35% answered that the place is suitable as it stands. One hundred per cent of visitors of all age answered that the place is full of green and open-air activities to do, making it very healthy as it stands. Of these respondents, 20% added that some views are very beautiful and give the perception of joy; these views include the sea, the mountains and the Vancouver skyline.

To the seventh question, "What is a healthy/happy/liveable place that you remember in this city or elsewhere?" all the locals answered False Creek and other places in Vancouver, including Coal Harbour and Capilano. Of visitors, 100%, in general, answered that this place is quite unique for the presence of many activities and the perception of liveability.

Finally, to the question, "How might the weather conditions influence the perception of liveability or happiness this place?" 100% of respondents – both locals and visitors of all age – answered that when the weather is too cold, the open places are less liveable, but there are many restaurants and pubs, Granville Island and a lot of sites in the area where it is possible to stay to enjoy the views or spend time.

## The transformation of the area

Regarding the traditional analysis, False Creek is an inlet located in downtown Vancouver and is accessible by public transportation or car. It is easily accessible for pedestrians as well. It is possible to take a little boat to cross some parts of the area.

Through to the 1950s, the False Creek area was the industrial heartland of Vancouver with sawmills and small port operations. After some years, the area was subjected to deterioration due to the change of the industry towards other areas. In 1960, the area experienced a large fire which destroyed part of the facility. Then, the future of False Creek was constructed by different debates concerning urban renewal. At the beginning, there was little consultation with people, but in 1968, with major public involvement started, it was established that an accessible waterfront with mixed-tenure housing and live-aboard marinas, a vibrant waterfront market, a wide park, communities centres and street-front shops and services would be created. In 1998, the Vancouver City Council adopted a set of blueway policies and guidelines, including the vision of a waterfront city which has to meet the environmental, cultural and economic needs of the city with sustainable, equitable and high-quality actions for people. In 2010, the Olympic Village, for athlete housing and logistics of the Winter Olympics, was built in Southeast False Creek, forecasting the development into a residential area with housing and services for about 11,000 to 13,000 people.

The current length of the seawall is 22 kilometres long, with an area present for pedestrians and cycle lanes in the two directions. The most important buildings include the Science World and the city's central stadium.

## The liveability and healthiness factors

False Creek is a place – like Coal Harbour – that resulted from a regeneration process initiated in Vancouver specifically devoted the be focused on "Liveability first".

People who use this place – due to the presence of many activities and facilities – are mainly locals, residents, professionals and visitors of all ages. Furthermore, people with disabilities have here comfortable paths and easy access.

The main factors which are reported in phase 5 of the method – the construction of the urban health, liveability and happiness map of analysis – are constituted by the general variety and beauty of the area, which include both natural and constructed parts; the park; the high-quality street furniture, the perception of safety and the playgrounds; the presence of the Granville market with the different activities and products; good accessibility; a wide and liveable waterfront; and the views on the sea, the skyscrapers, the bridges and the mountain.

All these elements create in this area a sort of continuum of liveable paths which is well used in terms of frequency and duration and whose health perception is felt in each part (Figure 22.20).

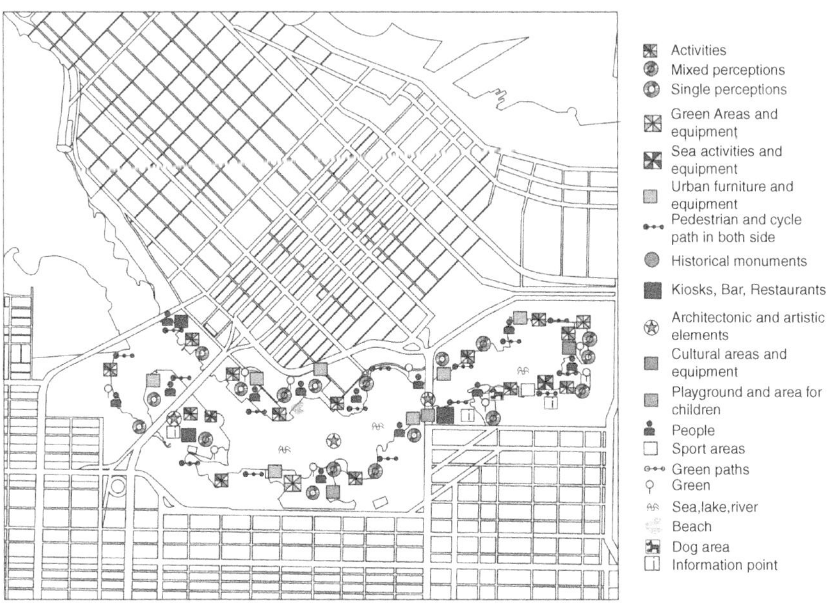


*Figure 22.20* Vancouver, False Creek, map of urban health and liveability analysis

## Designing health and liveability

Checking the consistency between the results of the different surveys and the Charter of 25 principles, in False Creek, nearly all the principles are respected.

The principles which could be improved in the area include (2) to encourage the use of the place by people of different age groups, from children to older adults; (7) to create suitable spaces for dogs and domestic animals; (9) to minimize or eliminate the noise generated by public transport; (15) to have the possibility of using the space in different weather conditions and seasons, contributing at the same time to its good state of maintenance; (16) to preserve both the place identity and the intangible characteristics of the site and its surroundings; (17) to both allow and promote different types of functions such as games, breaks, walking and so on; (18) to facilitate gymnastic activities with the presence of small equipment or a designated space; and (20) to encourage the presence of art in its different forms.

The demands of questionnaire which follow derive from the principles which could better satisfied.

1. What about creating new spaces or improve those already existing?
2. What do you think enhancing or creating connections between spaces in the False Creek area?
3. What about creating or improving the existing paths?
4. What do you think about increasing the inclusion of the green areas?
5. What about differentiating the activities for different age groups?
6. What about creating spaces for dogs and pets?
7. What do you think about creating light structures for different climatic conditions?
8. What about preserving the place's identity and the intangible cultural resources?

The questionnaire was administered to about 150 people aged between 25 and 70 years to both locals and visitors from all over the world. The language used was English. People were willing to give answers to the questions.

Regarding the first question, "What about creating new spaces or improving those already existing?" 80% of locals of all age answered that they agreed with respect to the parts of False Creek which are still in development or could be widened to create more open spaces. The other 20% answered that they agreed with this idea adding that they would prefer more spaces which have contact with the waterfront. With respect to visitors, 70% of all age answered positively, adding the importance of flexibility; the remaining 30% - mainly adults – answered that the place is already beautiful as it is and did not need other spaces.

With respect to the second question, "What do you think about enhancing or creating connections between spaces in the False Creek area?" 80% of respondents, both locals and visitors of all age, answered that they agreed. They added that the area is very big and that to go from a part to another, further connections could be useful. The remaining 20% of locals answered that they agreed with the connections between the different parts of False Creek, including between the Olympic Village and the rest of False Creek and between Grandville Island and the other parts of the site. The remaining 20% of visitors enjoyed the place and did not have a specific idea about it.

Regarding the third question, "What about creating or improving the existing paths?" about 70% of locals and visitors of all age answered that more than creating new ones, some paths could be further improved with respect to placemaking. The remaining 30% were also agreed, while locals added – mainly adults and older adults - that some paths need to be furtherly improved for people with disabilities and older adults, visitors suggested creating some little bridges.

To the question, "What do you think about increasing the inclusion of the green areas?" 80% of locals and tourists agreed with the idea and answered that in some parts, the green areas could be improved, including the Olympic Village and Granville Island. The other 20% agreed to improve the green areas, adding the importance of a good maintenance.

Regarding the fifth question, "What about differentiating the activities for different age groups?" 100% of those interviewed, both locals and visitors of all age, agreed with this idea, answering that it is important to have activities for children, young people, older adults and people with disabilities. Twenty-five per cent of locals added that many activities are suitable for people of all ages, but others require dedicated spaces.

To the question, "What about creating spaces for dogs and pets?" 80% of both locals and visitors agreed about creating off-leash areas where both dogs and pets can go free. The other 20% of locals agreed and added that in False Creek there are already some off-leash dog parks, but it is important to have more, also of small dimensions. The other 20% of visitors answered that off leash dog parks should be suitably equipped as well.

Regarding the question, "What do you think about creating light structures for different climatic conditions?" 100% of people, both locals and visitors of all age, agreed with this project idea. Of this, 20% of locals added that protection from the rainy and snowy weather is particularly needed.

With respect to the question, "What about the preservation of the place's identity and the intangible cultural resources?" 100% of locals independent of age answered that they agree, adding that cultural resources – including the environment and traditional activities such as traditional canoes – constitute an important element of False Creek. Of visitors, 30% agreed with the idea of preservation, and the other 70% agreed and would like to know more about its place identity and cultural resources.

Furthermore, research was carried out on social networks to understand the presence of this place on the internet and the degree of people's satisfaction. The Instagram profiles include False Creek Community Centre, Southeast False Creek Living and False Creek Ferries, with 1,195 followers. Hashtags include #falsecreekferries #falsecreekseawall #falsecreekvancouver and #falsecreek, with 125,000 posts. The Twitter profiles include @FalseCreekRes; @sefalsecreek and @FalseCreekFerry, with 2,812 followers. Facebook profiles include False Creek Community Centre; False Creek Residents Association and False Creek Co-op and Creekside Community Recreation Centre, with 1,209 followers. All this information and the thousands of likes on the posts confirm the liveability and the success of False Creek.

Information collected from internet websites such as TripAdvisor – with about 56,000 positive comments on this place and its attractions – or Booking.com – with 100 positive comments on the panorama, walkability, cyclability and the general beauty of these sites – are in line with the onsite questionnaire, although they are less specific with respect to the perceptions and information, such as the frequency of the activities which were carried out.

## Project interventions

The first project intervention is the creation of flexible spaces or enhancement of those already existing. The flexibility of the spaces is useful to have because places can fit different situations, including those of emergencies – environmental, pandemic-related and so on – or exceptional events of many kinds. The creation of new spaces or the enhancement of those already existing should be in line with these ideas.

The second project intervention is the enhancement or creation of connections between spaces. False Creek area is a place with many parts, including False Creek North, False Creek South, False Creek East, Olympic Village and Granville Island and each of them have further spaces within. To enhance the variety of spaces of False Creek, new connections, or the enhancement of those already existing, could be realized in relation to the internal parts and with the surroundings.

The third project intervention is the creation or improvement of paths. This project intervention concerns the creation of not only more cycle and pedestrian paths but also areas – where the space is of suitable wideness – for other sports and for people with disabilities. This intervention is in continuity with that of the creation of connections and should include the use of both sustainable material and attractive design.

The fourth project intervention is the enhancement of green space or its inclusion. False Creek has many parts with trees, gardens and parks. In some parts, in order to improve the attractivity of the paths greening, tree planting and vegetation in general could be carried out.

The fifth project intervention is the differentiation of activities for different age groups. In False Creek, it is possible to do many kinds of activities, including sport, walking and relaxing. To further improve its attractiveness, activities dedicated to children, young people, older adults and people with disabilities could be inserted.

The sixth project intervention is the creation of spaces for dogs and pets. Dogs and pets need off-leash spaces that are an important element for the liveability of an area both for the animals and their owners. The off-leash areas should comprehend equipment for games of different kinds in continuity with the surrounding natural environment (Figure 22.21).

The seventh project intervention is the creation of light structures for different climatic conditions. Rainy and snowy periods often make the open-air areas less liveable, decreasing their use. Suitable light structures, both to protect people from bad weather conditions and to allow them to practise some activities or to take a break, increase the health and liveability of the place. The structures should be mobile, built in natural or recyclable materials and designed with attention to the place's identity.

The eighth project intervention is the preservation of the identity of places and intangible cultural resources. Preserving the place's identity and the cultural resources is another important factor which can improve the liveability and healthiness of a place. In False Creek, there are many artworks as well as the natural environment; perceptions such as the visual ones, traditional sports and memory-making are fundamental elements of this space. To further improve its attractivity and liveability, public open spaces around them could be created or preserved or improved.

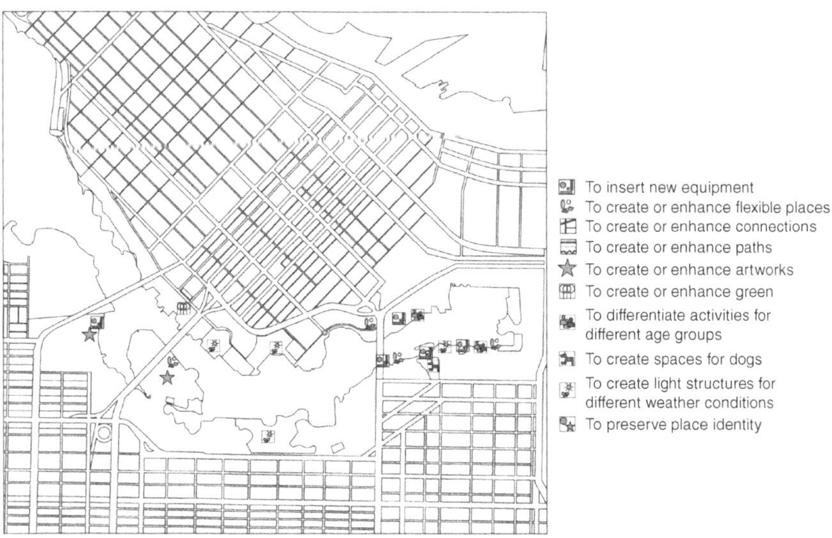


*Figure 22.21* Vancouver, False Creek, map of urban health and liveability design

## Conclusion

The two analysed areas are particularly emblematic because of the strong energies which the city of Vancouver is using to reach healthy, liveable and happy places for all. Coal Harbour and False Creek have characteristics and facilities which make them very comfortable places, which encourage their use for the many activities which offer, including taking a break. The suitable balance among a beautiful natural environment, the variety of activities which can be done – both free and for a fee – the good quality of street furniture and its good state of maintenance, the presence of art products which give a good and harmonic mix of new and old and the safety perception constitute some of the most important factors in Coal Harbour.

Furthermore, the particularly beautiful environment – composed by both sea and mountains – gives people the possibility of doing many things, at many times and for a long time, creating a satisfying experience which can be daily repeated.

The surveys and questionnaire demonstrate how a good urban design can be well perceived by people – both inhabitants and tourists – of different ages, cultures and habits and can contribute to its success in terms of socialization, liveability and sustainability. The information collected from internet websites – such as TripAdvisor and Booking.com – and the social networks – such as Facebook, Instagram and Twitter – confirm the beauty and success of these areas.

Then the proposed activities to improve Coal Harbour take into account some aspects that emerged during the surveys, questionnaires and the check of consistency with the Charter that could further improve its liveability. These include creating new spaces, creating spaces for different climatic conditions, inserting or enhancing artistic elements, differentiating activities for different age groups, inserting new equipment, creating suitable spaces for dogs and pets and reducing vehicle noise.

With respect to the False Creek area, the presence of the Olympic Village, the Science World and the Granville Island gives the area additional occasions of attraction, and in many parts of False Creek, there are sculptures of different types, which improve the perception of liveability and happiness often connected to the art.

The proposed activities to improve the area include creating flexible spaces or enhancing those that already exist, enhancing or creating connections between spaces, creating or improving paths, enhancing or inserting green areas, differentiating activities for different age groups, creating spaces for dogs and pets, inserting light structures for different climatic conditions and preserving the place's identity and the cultural resources.

Although some urban regeneration projects of both Coal Harbour and False Creek are still in progress, all the areas are designed with the health and liveability factors as a priority, connecting these to sustainability and high-quality design. The attention to people and their needs was in any case observed in the project realized until now making it an emblematic example of sustainable regeneration.

# 23 Wuhan

## The city

Wuhan is a city situated in the middle of China in the south-east of Hubei Province, known for its large number of lakes which characterize the territory.

The area which was chosen is the new Hankou riverfront of Wuhan. The first phase consists in the selection and definition of the study area. Three inspections on the new waterfront of Wuhan were carried out in different days and hours in order to delimitate the place for analysis. This area has a considerable size in terms of length and extension, but it was decided to consider the entire surface because, although it is characterized by different uses, it consists in a sort of *continuum*. Then, the boundaries which were determined concern all the area delimited by Yanijang Avenue, Wuhan Changjiang Tunnel and the cross between Yanijang Avenue and Chenjiaji Avenue.

The second phase is characterized by the observation of the characteristics of the place through three surveys concerning activities, perceptions and elements which contribute to the perception of liveability. The surveys begin by the part of the waterfront delimited by the Wuhan Changjiang Tunnel and continue towards the end part where Yanijang Avenue crosses Chenjiaji Avenue.

## People and activities

The first survey is devoted to the observation of the types of activities (enjoyment, passing by, work, etc.) and people (locals, visitors, professionals).

In the first part of the waterfront, the more frequent activities related to the port include maritime activities; walking, recreational, sport and leisure activities; and breaks. These activities are carried out by workers, traders who have their own kiosk or small business there or from street vendors who sell kites of considerable size and with a wide variety of colours and designs, locals with children and locals of different ages who play sports using the exercise equipment of the park, walk, fly the kites, sit on the steps and watch the river and the panorama. In the second part of the waterfront, the activities which are more present and frequent are strolling, sitting on benches

DOI: 10.4324/9781003098775-28

and enjoying the place. A less frequent activity is taking photos concerning one's own wedding and attending artistic performances. These activities are carried out by locals and Chinese tourists because of the presence of many hotels on Yanijang Avenue.

It can be deduced that the more present and frequent activities which make this place distinctive in the waterfront as a whole include flying kites, practised by adults and children, which is a very dynamic and choreographic activity; practising sports exercises of various types; taking breaks in the park on the tiers or on benches; and strolling among sculptures and trees.

The second survey concerns single and mixed perceptions. Single perceptions which are most present are the visual, sound, smell and ones touch. The visual perceptions include the perception of the kites (high and pleasant), the landscape observed from the tiers (high and pleasant), perspective of the cherry trees (medium and pleasant) and the historic buildings of the European Concession (medium and pleasant). The sound perceptions include the sound of silence, which prevails in some places; the wind; and the voices and movements of children who play in the final stretch of the waterfront. The smell perception concerns the presence of various types of vegetation and trees in the gardens (high and pleasant). The tactile perception concerns the sculptures which invite socialization as well, such as that of children playing in a circle and that of the globe, in both cases high and pleasant, and the perception given by the sand present in the initial stretch, high and pleasant (Figures 23.1–23.12).

*Figure 23.1* Wuhan, Hankou waterfront, the path

*Figure 23.2* Wuhan, Hankou waterfront, the beach and people on the stairs

*Figure 23.3* Wuhan, Hankou waterfront, engaging sculptures

*Figure 23.4* Wuhan, Hankou waterfront, the flying kites

*Figure 23.5* Wuhan, Hankou waterfront, perspective

*Figure 23.6* Wuhan, Hankou waterfront, perspective with the open-air structure for events

*Figure 23.7* Wuhan, Hankou waterfront, perspective with a gazebo

*Figure 23.8* Wuhan, Hankou waterfront, perspective with a tree path

*Figure 23.9* Wuhan, Hankou waterfront, perspective with shelters

*Figure 23.10* Wuhan, Hankou waterfront, perspective sculptures

*Figure 23.11* Wuhan, Hankou waterfront, the garden with water games

*Figure 23.12* Wuhan, Hankou waterfront, the gymnastic equipment area

Regarding the mixed perceptions, those which are more present include serenity and joy, especially in areas where people play with the kites or admire the gardens, in both cases high and pleasant, and of light melancholy, where people look at the great ships of the port, medium and non-influential.

The pace of these activities is generally moderate or slow, providing a very pleasant perception of this place.

The third survey concerns feature which contribute to the perception of liveability. The elements of the park contributing to this perception include the part of the park with sand; the garden areas with the presence of water features, green areas and cheery trees; and the sculptures of human scale that mime different and aggregating actions of people.

## The places from peoples' points of view

The third phase concerns the questionnaire to users of the place, aimed at understanding the elements which give them here the perception of liveability, healthiness and happiness.

The questionnaire was administered in English to approximately 100 people, tourists – mainly from China, the UK and the US – and passers-by.

Questions which were administered are the following:

1. Does this place give you a feeling of happiness or sadness/liveability or discomfort/health or unhealthy?
2. What are the elements that give you the above sensations?

3. What are the main facilities that give quality to this place?
4. What are the activities that you do in this place, and how often?
5. What do you think about the presence of many or few people here? Do you think that it is capable to improve the pleasantness or unpleasantness of the place?
6. According to the place's healthiness, liveability and happiness what could be done in order to improve this place?
7. What is a healthy/happy/liveable place that you remember in this city or elsewhere?
8. How might the weather condition influence the perception of the health, liveability or happiness of this place?

To the first question, "Does this place gives you a feeling of happiness or sadness/liveability or discomfort/health or unhealthy?" 100% of those interviewed, independent of age and nationality, answered liveability.

As to the second question, "What are the elements that give you the above sensations?" the answers were very different from each other. Thirty per cent of people of all ages and nationalities answered the concession buildings behind the waterfront area; 40%, the river, the cheery trees and nature in general, and the park in its entirety; and 30%, the sculptures.

As to the third question, "What are the main facilities that give quality to this place?" 30% of all people answered benches and tiers, 30% answered the equipment for gymnastic activities and the remaining 40% answered the picnic gazebos, the small structures to shelter people from the sun or rain and outdoor infrastructures for musical events.

To the fourth question, "What kind of activities do you act in this place, and how often?" the activities mentioned were very different, probably depending on the variety of them which are possible in this area. Of both locals and visitors, 80% answered that the activities which are carried out more frequently include walking in the green areas, taking breaks on the tiers, kite-flying and practising sport with the equipment for gymnastics or music. The remaining 20% answered observing the river, taking the marriage photos and eating or drinking sitting in the gazebo.

To the fifth question, "What do you think about the presence of many or few people here? Do you think that it is capable to improve the pleasantness or unpleasantness of the place?" 100% answered that the presence of many persons improves the pleasantness of the place.

To the sixth question, "According to the place's health, liveability and happiness, what could be done in order to improve this place?" 100% of people answered nothing, because they like and enjoy this area. Of this, 20% added inserting elements capable of giving a greater distinctiveness to this place in accordance with the historic concession area which stands behind it.

As to the seventh question, "What is a healthy/happy/liveable place that you remember in this city or elsewhere?" 100% of the local people answered with different places in Wuhan, from the Yellow Crane (Huanghe Lou) to the east of the park lake, also including the Hankou waterfront; visitors answered many places, all of the Hankou waterfront.

Finally, to the question, "How might the weather condition influence the perception of health, liveability or happiness of this place?" 100% of people answered that in the rainy months, that place is less used than the other months because is an outdoor area. However, the presence of shelters and areas for the sport and for children make this place useable also in the cold months. Furthermore, during festivals and special events, Hankou is well used in all seasons because of the presence of equipment for artistic performances.

## The transformation of the area

The fourth phase concerns the analysis of the traditional cartography at urban scale. Wuhan is a city situated in the south-east of Hubei Province, known for its lakes and related landscape which characterize the whole territory. Wuhan has been involved in three important floods in the 20th century – 1931, 1954 and 1998 – with great destruction in terms both of the built and the natural environment.

The development of the Hankou waterfront started after the last flood, in 1999, with the function of creating both flood control and an urban design of quality.

The waterfront area of analysis is about 4 kilometres and, as mentioned before, is located between Yanijang Avenue, Yangtze River which runs along this area, Wuhan Changjiang Tunnel and the cross between Yanijang Avenue and Chenjiaji Avenue (Grove and Zhang 2020; Law and Qin, 2013; Liu et al., 2004; Liu, 2012; Meng and Liu, 2007; Shu and Qin, 2014).

The waterfront area has a linear plan constituted both from green and built areas. The greater particularity of this site is that it borders five concession areas: French, English, German, Russian and Japanese. This particularity gives to this place a variety of architectural styles as well as the different bridges which cross the riverfront.

## The liveability and healthiness factors

The fifth phase concerns the identification of the elements that contribute to the liveability and healthiness of this place. The liveability of Hankou Riverfront is given by the variety of functions that this place offers because it constitutes a suitable pause to the close traffic road. It is also a place full of history, witnessed by the buildings of the five concessions situated behind it.

Along several stretches of the waterfront, it is possible to enjoy nature – with its perfumes and colours – the water games and the sculptures; sit on the benches or the steps; and look the river.

The site also offers the possibility of dynamic activities such as practising sports with the exercise equipment, flying kites, walking on the sand and letting the children enjoying the playground. Furthermore, the equipment

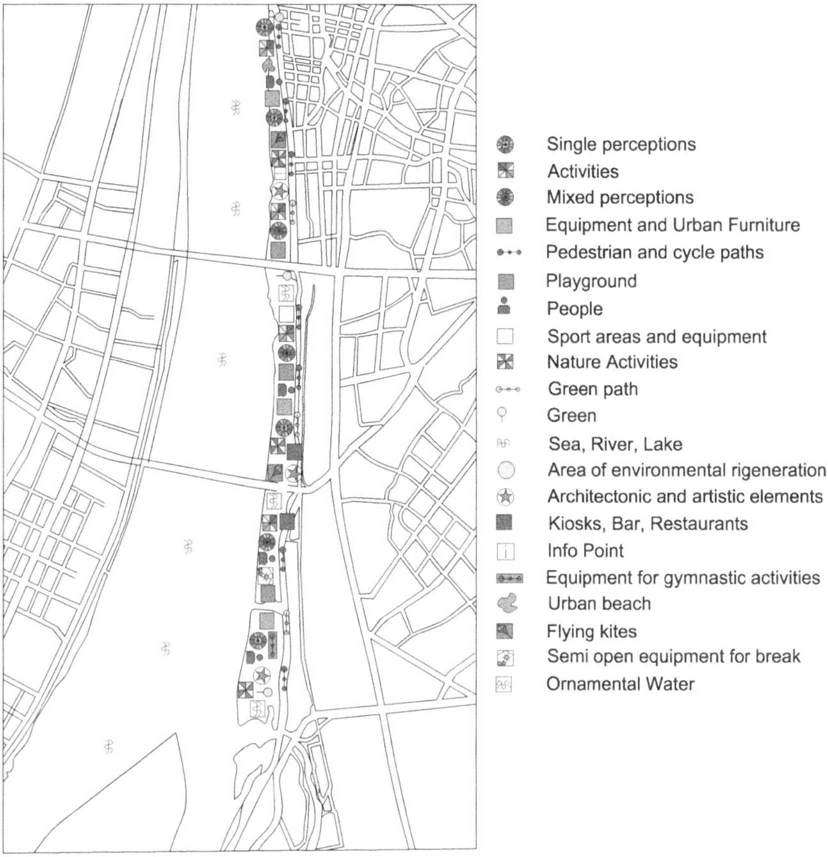


*Figure 23.13* Wuhan, Hankou waterfront, map of urban health and liveability analysis

for events, such as the amphitheatre, gives the possibility of attending artistic performances during festivities.

The sculptures are of particular interest, involving visitors and attracting them to this place. As the answers to the questionnaire have shown, at the new waterfront of Wuhan, the activities are also the most frequent and those that offer the greatest feeling of liveability to people.

The success of this area stands in its uniqueness for the city of Wuhan, as there exists a right balance between nature and landscape, games and sports activities and the possibility of taking a break. Furthermore, the park is built with quality materials and accurate design, despite a preponderant use of cement, offering characteristics of beauty as well (Figure 23.13).

## Designing health and liveability

Checking the degree of health, liveability and happiness is carried out through the 25 principles of urban health and liveable design concerning the Charter. By checking the principles, the area is in line with almost all. In all its extensions, as it has emerged by the previous phases, the path contains a wide variety of elements which makes it healthy, liveable and happy. The principles which could be improved include (7) to create suitable spaces for dogs and domestic animals; (9) to minimize or eliminate the noise generated by public transport; (15) to have the possibility of using the space in different weather conditions and seasons, contributing at the same time to its good state of maintenance; (16) to preserve both the place identity and the intangible characteristics of the site and its surroundings; and (24) to promote the educational function which a place has – for example clearly displayed information about history of the place and so on or suitable ways to use it – increasing its intrinsic value.

The items of the questionnaire which follow derive from the principles which could better satisfied.

The questionnaire was administered to about 100 people aged between 25 and 70 years, mainly locals and visitors from China, the UK and the US. The language used was English, and people were willing to give answers to the questions.

The following questions, and their answers, are reported.

1. Do you use this area during all periods of the year? What do you think about creating urban public light structures to use this place also in periods with bad weather conditions?
2. What about inserting old or contemporary art products in their different forms to contribute to the sense of liveability and happiness of this place?
3. What do you think about widening the game areas by inserting more playgrounds and places for games for people with disabilities and older people?
4. What do you think about inserting more information points?
5. What about widening the use of new technologies to give a more profound experience of the place?
6. What about minimizing the sound of vehicles in the parts where this is more evident?
7. What about creating an area devoted to dogs?

As regards the first question, "Do you use this area during all periods of the year? What do you think to create urban public light structures to use this place also in periods with bad weather conditions?" 80% of locals of all ages answered that they use always this area, although during periods of bad weather they use it for less time. The other 20% answered that they do not go to the waterfront when it is too cold or warm or when there is flooding. One hundred per cent of those interviewed, however, observed that some shelters already exist but agreed with the insertion of further light structures to repair from temperatures that are too cold or warm.

With respect to the second question, "What about inserting old or contemporary art products in its different forms to contribute to the sense of liveability and happiness of this place?"

Of both locals and visitors, 100% agreed with this idea. Thirty per cent added that in many parts of the park, there are already sculptures, but they agreed to add different kinds of art products in other parts of the place.

With respect to the question, "What do you think about widening the game areas or inserting more playgrounds and places for games for people with disabilities and older adults?" 80% of locals and visitors answered that there are many playgrounds for children and many exercise equipment for people of all ages. Then they agreed about adding playground areas for people with disabilities and older people and exercise equipment for people with disabilities. The other 20% of locals agreed with adding more playgrounds and games for all.

Regarding the question, "What do you think about inserting more information points (such as totems, maps and so on) with the process of regeneration of this place?" 90% of local people and visitors who were interviewed, independent of the age, answered that they agree to insert totems with maps and more generally information points. Locals and visitors are interested to know more information concerning the place, the equipment, the events and so on. The other 10% of locals answered that they come in this waterfront only to enjoy the landscape and to walk or take a break; they know this place well and do not need more information.

To the question, "What about widening the use of new technologies to give a more profound experience of the place (e.g. virtual maps, virtual experience connected to the history and local products)?" the answers to this question were connected to the previous one. The area is very extended and with different characteristics; furthermore, there are many historical buildings located in the surroundings. One hundred per cent of locals and visitors are interested in having a more profound experience of the place, which is also related to local products, but while young people and adults are more interested in the virtual tools, older adults prefer totems or other kinds of "physical" information points.

To the question, "What about minimizing the sound of vehicles in the parts where this is more evident?" 100% of locals and visitors agreed with this idea, adding that this is needed only in some parts of the area because in many parts the sound of vehicles is not perceived.

Finally, to the question, "What about creating an area devoted to dogs?" 80% of people, in particular locals who live close to the park, agreed with this idea because in this way, dogs, with their owners, can freely run in an area specifically dedicated to them. The other 20% agreed and added that this area should have also a dedicated access point.

Checking social networks and websites with suggestions, no official Instagram page is present, but there are pages using the hashtag #hankou, with 5,359 posts.

There are no official pages on Facebook, but there are some pages about Wuhan that show photos of the waterfront with positive comments on the beauty of the walk, the sculptures and the beach, and there are many movies on YouTube which also show images of the waterfront.

Among the travel websites, on TripAdvisor, there is the presence of Beach Park, as the first of the Wuhan parks, with about 106 comments. These concern the variety of paths; the beauty of nature, in particular, of the cherry trees in bloom; and the activities that can be carried out in the park. (Figure 23.14).

From the information gathered in the previous phases, the following project interventions were designed.

The first intervention concerns creating new spaces with artistic objects. As emerged in the previous phases, the creation or even the expansion of some spaces dedicated to artistic products also for the purpose of the game can help expand the liveability and urban happiness of the place. Everything should be done in continuity with the artistic elements already present along the waterfront.

The second intervention is the enhancement of the identity of the places. The waterfront area has many historic buildings of the European Concession on the outside. Within the area, there is no reference to this historical part

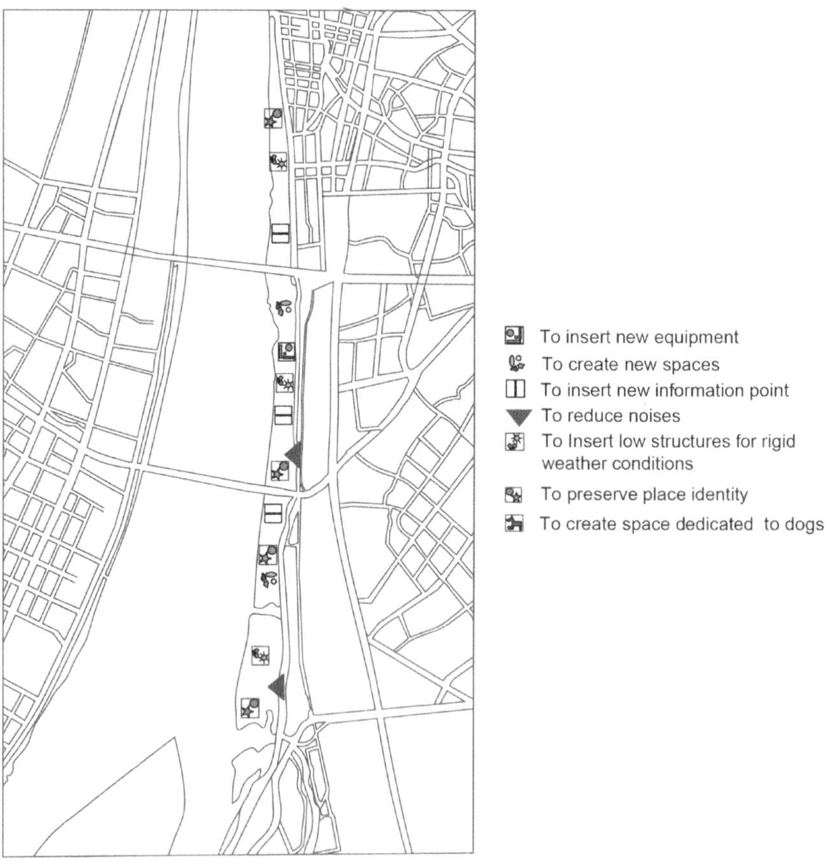


*Figure 23.14* Wuhan, Hankou waterfront, map of urban health and liveability design

which is visually perceived only from some points. For a greater enhancement of the identity of the place, some spaces could be improved to allow a more direct visual corridor together with specific street furniture. Information points could also be inserted to describe the historic buildings.

The third intervention is creating suitable spaces for the different climatic conditions. The waterfront of Hankou is widely used as a place for leisure thanks to the variety of spaces and the considerable facilities present. It is also used in periods of flooding, albeit mainly by young people. During periods of more rain or of very hot or cold weather, the place is still used but for a shorter duration. The intervention is therefore aimed at creating spaces and/or equipment suitable for greater resilience to those climatic conditions. These are to be built in continuity with the identity of the place and to be mainly mobile and flexible.

The fourth intervention concerns inserting new equipment. As emerged from the previous phases of the method, there are already equipments for game and motor activities, but these are mainly for people with good motor skills. It is therefore possible to provide equipment for soft physical activity and for people with disabilities, also with dedicated perceptive paths.

The fifth intervention is inserting information points along the waterfront. The information points should be of two types, with the possibility of networking (e.g. QR codes, multimedia totems) and illustrative through plans and different types of data on the place – historical and gastronomic as well as information related to routes, present equipment, timetables for means of transport and more. As with all the elements of the park, these should be thought of in continuity with the identity of the place.

The sixth intervention consists in reducing the noise coming from the means of transport. The waterfront is bordered by a driveway which, in some parts of the park, causes noise and can decrease the feeling of healthiness and liveability of this place. In order to create an adequate acoustic barrier, in those points of greatest noise from vehicular traffic, the intervention proposes to insert a green filter with the addition of trees, hedges and other types of natural elements aimed at having a greater barrier to noise and offer further liveability to the users of the place.

## Conclusion

This chapter has, in particular, illustrated the case of the new Hankou riverfront of Wuhan, China – carried out in the framework of the FP7-PEOPLE-2011-IRSES Project Planning, Urban Management and Heritage – PUMAH and the CNR-CACH Italy–China Bilateral Agreement 'Cultural Itineraries and the One Belt One Road Initiative. Good practices and policies for conservation and tourism development', which has been built since 1999 both for flood defence reasons and to create a public space of quality which borders the historic concession areas.

Following the results from the Healthy Pl@ce Design method, the urban health and liveability of the Hankou riverfront is given by the variety of functions that this place offers because it is a suitable pause to the near road traffic. It is also a place full of history, witnessed by the buildings of the five concessions situated behind. In several stretches of the waterfront, it is possible enjoying the nature – with its perfumes and colours, the water games and the sculptures, sitting down on the benches or on the steps and looking the river. The site also offers the possibility of dynamic activities, such as practising sports with the exercise equipment, flying kites, walking on the sand and letting children enjoy on playground. The equipment for events such as the amphitheatre gives the possibility of attending artistic performances during festivities.

The sculptures are of particular interest, involving visitors and attracting them to this place. As the answers to the questionnaire have shown, in the new waterfront of Wuhan, the aforementioned activities are also the most frequent and those that offer the greatest feeling of liveability to people. The success between this area stands in its uniqueness for the city of Wuhan, as there exists the right balance of nature and landscape, games and sports activities and the possibility between taking a break. Furthermore, the park is built with quality materials and accurate design, despite a preponderant use of cement, offering characteristics of beauty as well.

The project interventions emerged by the results of the phases of analysis and design include creating new spaces with artistic objects, enhancing the identity of the places, creating suitable spaces for the different climatic conditions, inserting new equipment, inserting information points along the waterfront and reducing the noises coming from the means of transport.

Wuhan is a constantly evolving megacity, and this waterfront is one of the emblematic areas of this transformation. The identified project interventions are aimed at a further definition of this area in relation to the health and liveability of the place and concern both the tangible and intangible aspects in an idea of a connection with the adjacent concession area, on one side, and of attention to the environment, on the other.

# Conclusion

## Principles of Urban Health and Liveability Design

This book has dealt with issues relating to sustainable urban regeneration from the point of view of urban health and liveability. After the illustration of the principles of the New Urban Agenda and the 17 Sustainable Development Goals, the topics investigated concerned resilience and adaptation, mixed connections, healthy urban places and indicators.

The 2030 Agenda goals, as well as the principles concerning the 2016 New Urban Agenda, present a holistic approach; indeed, topics such as resilience, equity, health, safety, sustainable economic growth, resilient infrastructures and nutrition improvement are some of those in common between the two Agendas and constitute the wider framework of this book as well.

Resilience and adaptation are to be considered fundamental in dynamic planning, where the uncertain and the sudden become components of the planning process with the same weight as the others and where urban planning tools are renewed or modified in order to contribute to the management of any crisis in an appropriate way. While it is true that unforeseen events, by definition, occur suddenly, it is also true that adaptive and dynamic planning can better support the creation of a new equilibrium.

The issues inherent in healthy places are many, with a lot of interrelationships and should be treated with strategies at different urban scales to promote healthier places and behaviours. The COVID-19 pandemic has underlined even more the relationships between elements on a global scale and factors on a local scale, between the quality of places and the quality of life of people, highlighting the importance of open public spaces both as spaces in themselves and as spaces of relationships between people and built environments.

In this context, multiple but sustainable connections between different kinds of infrastructures, both tangible and intangible, should be realized to improve interaction, inclusiveness and equity.

The methodological approaches chosen to analyse the topics of the book included the 15-Minute (Moreno, 2020), flexible (Bergevoet and Van Tuijl, 2016), soft (Sim, 2019), smart (Sassen and Kourtit, 2021) city and the Healthy Pl@ce Design (Sepe, 2020a).

The rediscovery of proximity translates into resilience in the face of crises that can manifest itself over time (environmental, pandemic, etc.),

DOI: 10.4324/9781003098775-29

moving towards the principles of sustainable development and the protection of urban health. A polycentric life means urban and territorial resilience with cities within a short distance at all territorial scales. Flexibility constitutes an objective as well as a methodology to aim for in order to achieve sustainability – meant in its triple meaning – and reduction of risks. Another aspect to take into account is the attention to the identity of the places. Namely, considering the different variables that can occur during a process of planning and designing a place, some constants, such as the identity of the place, must be taken into account in order to protect and enhance it. The concept of a soft city can also be extended to the use of technologies and where different types of information can be offered to citizens in real time. The design of soft places in this sense never ends as communities change, as do buildings and their public spaces, physical and digital, as well as their uses and habits. Finally, the potential of smart places is vast, and in order to be used in a sustainable way and be of real help for urban health and liveability, it is necessary that the new technologies are more closely connected with the peculiarities of the places and the desires and needs of citizens so that real and virtual spaces can collaborate in synergy and technologies would represent an added value to the transformation of cities and concrete support in times of crisis.

Particular attention was paid in the book to the original Healthy Pl@ce Design Method which was explained in detail together with the features of the supporting software and indices. This method is both rigid and flexible as it consists of a pre-established investigation protocol that is repeatable and can be easily adapted to be used in different types of contexts and for different objectives.

This theoretical part was the setting for the empirical part relating to the case studies carried out in public spaces of Europe, Canada and China – Bordeaux, Copenhagen, Hamburg, Madrid, Newcastle–Gateshead, Nice, Dublin, Vancouver and Wuhan – particularly emblematic for the cities to which they belong.

These experiments allowed the identification of the principles of the Charter of Urban Health and Liveability Design, illustrated in the following. These principles are thought with a holistic approach and contain the topics shown in the previous chapters of the book, regarding both theoretical and empirical parts (Lehmann, 2010; Xu, 2011; Sepe, 2016, 2022).

1. *A healthy and liveable place is a space which can transmit feelings of health and liveability to everyone who uses it*

   Accordingly, it is important to do the following:
2. *To encourage the use of the place by people of different age groups, from children to the older people*

   It is important that people of all ages can use public spaces; to allow this, different kinds of paths, equipment and activities should be provided. The possibility of carrying out activities devoted to all ages gives variety, increasing the attractiveness of the place.

3. *To eliminate architectural barriers which might discourage people from frequenting that space*
   More generations of people living together constitute a factor of importance because it puts into evidence the presence of different activities, times and rhythms of diverse use and the lack of architectonic barriers, which can constitute obstacles for children, older adults and people with disabilities. It is important to eliminate any kind of architectonic barriers to promote a wholly inclusive use of places.
4. *To create a suitable balance between the elements of nature, landscape and equipment in the composition elements of the space*
   The environment in the different forms, as those generated by nature and those generated by humans (landscape) are important to allow sensorial activities such as view, touch and taste. A suitable mix of environment and equipment to enjoy them constitutes an important factor for the urban health and liveability of a place.
5. *To have both in streets and public spaces natural lighting during the day and artificial lighting at other times, avoiding artificial light in daily hours*
   The natural and artificial light constitutes an important factor for urban health because it can be both associated with different actions and allow a condition of safety. Sunlight allows conditions for both physical and emotive safety and well-being. Furthermore, the artificial light allows many actions such as walking, biking, taking a break and other actions which cannot be easily carried out during nighttime.
6. *To retain an adequate state of cleanliness and maintenance*
   Managing a public space, with suitable care to its state of cleanness and maintaining its different parts, is a necessary condition for determining a factor of well-being in that place. Poor maintenance offers an image of disuse which can favour non-positive behaviours.
7. *To create suitable spaces for dogs*
   People are often accompanied by dogs, but there is no suitable space for them. Creating a space where dogs can run, play and rest is important for the well-being of both animals and their owners.
8. *To create a sense of security and safety to those who walk, cross, rest and do different activities in the public spaces*
   The presence of functions which can be carried out during the day contributes to the creation of a perception of safeness because it offers the idea of living in a place with a moderate pace. The presence of those actions has to be joined to suitable lighting – as well as with artistic function – in order to improve the perception of safeness.
9. *To minimize or eliminate the noise generated by public transport*
   The sound perceptions can be constituted by elements capable to offer perceptions of health and liveability, such as in the case of the sound of the elements of nature, namely not only that of water but also that of vegetation in its different shapes. The vegetation can also constitute an element to create acoustic barriers to the means of transport,

realizing an adequate separation between a place of quick crossing and a space for taking a break and free time. The separation from the car traffic sounds can be also carried out forecasting other kinds of filters such as cycling lines which work as a place between space for breaks – with quiet pace – and that of car movement – necessarily quicker.

10. *To improve suitable pedestrian and cycle lanes*

    Many streets are created only or mainly for cars and motorized vehicles. It is necessary for the liveability, healthiness and happiness of places to introduce or improve pedestrian and cycle lanes which are safe and suitably designed. These should be created in a logic of integration and network with the other infrastructures of different kinds of the city.

11. *To fully perceive naturally occurring smells – for example wood, grass, sea.*

    Beyond the sound and visual perceptions, the smell ones are also important for urban liveability. The smell of nature, both vegetal and flower, has an important function for emotivity. Furthermore, the use of natural materials and the realization of urban equipment and furniture contribute to the solicitation of both smell perception and well-being.

12. *To have direct contact with natural materials, preferably local, used in the design of the space*

    The use of natural material for urban furniture, equipment and paving offers the possibility of creating diverse kinds of perceptions in the people who use it. When the material is local, this has the double effect of contributing to the maintenance of place identity, favouring the maintenance and management of the place as well.

13. *To have the presence of water in different shapes which promotes the vitality of the place*

    Water, in its shapes and sounds (e.g. fountains, paving with water games, etc.), can create aggregative and socialization activities and constitute factors which are important for creating emotions in a place.

14. *To have the possibility of doing actions – such as walking, watching and so on – with a moderate or slow pace to promote opportunities to take breaks in the space*

    The pace imposed by contemporary society is often quick. Many actions are carried out simultaneously also with the use of new technologies which favour this behaviour. The pace can be reduced with the presence of suitable break spaces and places for enjoying both natural and artistic elements.

15. *To have the possibility of using the space in different weather conditions and seasons, contributing at the same time to its good state of maintenance*

    A place should be thought about as useable in all seasons, guaranteeing suitable maintenance in times of less light and months with lower or higher temperatures. The presence of light structures for both breaks and free time also in continuity with other structures which are external of the public space would contribute to its adaptive use and maintenance during all the year.

16. *To preserve both the place's identity and the intangible characteristics of the site and its surroundings*

    The identity of a place, meant as the set of characteristics which make the place unique and recognisable, is given by the urban and architectonic elements – namely the tangible ones – and by elements connected to the perceptions – namely the intangible ones. A smell or a view is capable of making people remember a place or a square or an architecture and determine positive emotions in those who live near them. Furthermore, it is important that the surroundings are agreeable and enhance the identity of place as well in order to create a place which is suitably harmonized with its external parts.
17. *To both allow and promote different types of functions such as games, breaks and walks*

    It is important that in a place, different kinds of functions are provided together with the possibility to add new ones in accordance with new needs – such as those related to a pandemic or other kinds of emergencies – with both a resilient and a flexible approach. Furthermore, different activities allow a mix of uses in various periods of the year.
18. *To facilitate gymnastic activities with the presence of small equipment or a designated space*

    The presence of sport equipment constitutes an element capable to make more attractive and healthier the place; this should be suitable for all people, including older adults and those with disabilities.
19. *To have the possibility of doing actions that normally are not permitted – such as walking barefoot in the water or in designated public areas – improving a feeling of freedom and joy*

    A place such as a public space is a site where it is possible to imagine activities which usually are not carried out in other typologies of places. The presence of materials such as sand, water or others which allow actions, such as sitting down without shoes, that offer a perception of freedom and well-being.
20. *To encourage the presence of art in its different forms*

    Art constitutes a factor able to cause emotions favouring a condition of liveability. Furthermore, some artworks constitute aggregative elements themselves due to the potential of dynamism which they possess.
21. *To promote sculptures, games or other elements and amenities which can bring a smile to a person's face that promotes a state of liveability and happiness*

    There are many activities which vitalize a place and people mutually, creating a perception of happiness and well-being in people who take that action and in those who observe it. The typical case in this sense are the installations or the "engaging" sculptures which invite to interact with them. These include the creation of shadows by people using artificial light reflected on a wall, the possibility to create little green spaces, making games through water paving, drawing on wall and panels suitably posed for this function and so on. Furthermore, the possibility of live installations and sculptures, with their urban value, favours the atmosphere of happiness of the place.

22. *To promote participation, namely the feeling of being able to contribute to the life of that place increasing the sense of belonging*

    Organizing focus groups on specific topics related to the public space or the neighbourhood is important to those participating in the life of the place and contributes to its suitable maintenance, improving its liveability and, at the same time, the sense of belonging.

23. *The consideration of the place as a symbolic of the neighbourhood improves the perception of its identity*

    A place become symbolic when people for different reasons make it their own, achieving a sense of belonging. The presence of elements of the tangible and intangible culture strongly contribute to the process of recognizability. In this way, the place's identity is powered, and its perception by people is enhanced.

24. *To promote the educational function which a place has – for example clearly displayed information about the history of the place or suitable ways to use it – increasing its intrinsic value*

    The educative function of a place can be exercised in different ways. Attention to care of the environment is one of the more relevant modalities. Other examples include realizing little green space with didactic functions for apprehending the life cycle of products and the importance for the environment, realizing little water mirrors with comprehension of the cycle of the hydric resource and more. Every place can suggest, according to its peculiarities, a different example, which can include respecting particular categories of people, such as older adults, children and people with disabilities, in an equity logic.

25. *To facilitate the use of new technology to increase the knowledge of its intangible values and history, offering a more profound experience of the place*

    The new technologies for internet constitute adding value if they are used to support the knowledge of a place and its use. The presence of wireless, multimedia totems and QR codes constitute elements which can support the knowledge aspect of the place in its different aspects and can be designed in a logic of integration with the other infrastructures and networks of the city. Furthermore, these can constitute reasons for major use, socialization, sustainable economic growth and success.

The principles illustrated earlier should be considered dynamic, in keeping with the increasingly rapid rates of change in a place, which lead to continually updating the needs of people with respect to the healthiness and liveability of a place. These can be considered both as a checklist and as guidelines for sustainable urban places, which have flexibility, resilience, adaptation, integration, participation and inclusion among the main factors to take in account.

# References


AAVV (2011), *Copenhagen Climate Adaptation Plan*, Copenhagen: City of Copenhagen.

AAVV (2012), *The CPH 2025 Climate Plan*, The City of Copenhagen, Technical and Environmental Administration, Copenhagen.

AAVV (2018a), *Smart Placemaking: Future Healthy Communities*, https://placetech.net/insight/smart-placemaking-future-healthy-communities/

AAVV (2018b), Copenhagen: resilience and liveability, Field Actions Science Reports Special Issue 18, http://journals.openedition.org/factsreports/4750

AAVV (2019), *Adoption of CPH Municipal Plan 2019*, https://livplan.eu/adoption-of-cph-municipal-plan-2019

Abajo, L. L. D., Pérez-Fortes, A. P., Alberti, M. G., Gálvez, J. C., Ripa, T. (2020), 'Sustainability analysis of the M-30 Madrid Tunnels and Madrid Río after 14 years of service life', *Applied Sciences*, 10(20): 7368.

Adams, D., Tiesdell, S. (2013), *Shaping Places: Urban Planning, Design and Development*, London: Routledge.

Ajuntament de Barcelona (2013), *Urban Mobility Plan of Barcelona 2013–2016*, Barcelona: Ajuntament de Barcelona.

Anru et al. (2007), *Projet Urbaine du quartier de l'Ariane. Phase I. Convention*, Nice Cote D'Azur.

Appleton, J. (1975), *The Experience of Landscape*, New York: John Wiley.

Appleyard, D. (1981), *Liveable Streets*, Berkeley: University of California Press.

Astleithner, F., Hamedinger, A., Holman, N., and Rydin, Y. (2004), Institutions and indicators: The discourse about indicators in the context of sustainability, *Journal of Housing and the Built Environment*, 19(1): 7–24.

Badland, H., Whitzman, C., Lowe, M., Davern, M., Aye, L., Butterworth, I., Hes, D., Giles-Corti, B. (2014), 'Urban liveability: Emerging lessons from Australia for exploring the potential for indicators to measure the social determinants of health', *Social Science and Medicine*, 111: 64–73.

Ballas, D. (2013), What makes a 'happy city'?, *Cities* 32: 539–550.

Banica, A., Kourtit, K., Nijkamp, P. (2020), Natural disasters as a development opportunity: A spatial economic resilience interpretation, *Review of Regional Research*, 40: 223–249.

Barton, H. (2016), *City of Well-Being: A Radical Guide to Planning*, London: Routledge.

Benedict, M. A., McMahon, E. T. (2006), *Green Infrastructure: Linking Landscapes And Communities* Washington DC: Island Press.

Bergevoet, T., Van Tuijl, M. (2016), *The Flexible City: Sustainable Solutions for a Europe in Transition*, Rotterdam: Nai010 Publishers.

Bohland, J. R., Davoudi, S., Lawrence, J. L. (eds.) (2019), *The Resilience Machine*, New York: Routledge.

Bonnet-Galzy, M. C. (2014), *Observatoire national de ZUS*, Rapport 2014.

Borsekova, K., Nijkamp, P. (2019), *Resilience and Urban Disasters*, Cheltenham, UK: Edward Elgar Publishing.

Boyko, C. T., Gaterell, M. R., Barber, A. R. G., Brown, J., Bryson, J. R., Butler, D., Rogers, C. D. F. (2012), Benchmarking sustainability in cities: The role of indicators and future scenarios, *Global Environmental Change-Human and Policy Dimensions, 22*(1): 245–254.

Brinkerhoff, J. (2002), Government-nonprofit partnership: A defining framework, *Public Administration and Development, 22*(1): 19–30.

Brown, J., Barber, A. (2012), Social infrastructure and sustainable urban communities, *Proceedings of the Institution of Civil Engineers-Engineering Sustainability*, 165(1): 99–109.

Brown, L. (1994), 'The seven deadly sins of the information age', *Intermedia* 22(3): 32–37.

Brownson, R. C., Baker, E. A., Housemann, R. A., Brennan, L. K., Bacak, S. J. (2001), 'Environmental and policy determinants of physical activity in the United States', *American Journal of Public Health*, 91: 1995–2003.

Bulkley, K. E., Henig, Jeffrey R., Levin, H. M. (eds.) (2010), *Between Public and Private: Politics, Governance, and the New Portfolio Models for Urban School Reform*, Cambridge, MA: Harvard Education Press.

Burgos, F., Garrido, G., Porras-Isla, F. (2014), *Landscapes in the City. Madrid Río: Geography, Infrastructure, and Public Space*, Madrid: Turner.

Burns, G. W. (2005), Naturally happy, naturally healthy: The role of natural environment in well-being. In F. A. Huppert, N. Baylis, B. Keverne (eds.) *The Science of Well-Being*, New York: Oxford University Press.

Burton, L. (2015), Mental well-being. In H. Barton, S. Thompson, S. Burgess, M. Grant (eds.) *The Routledge Handbook for Health and Well-Being*. London: Routledge.

Buttimer, A., Seamon, D. (1980), *The Human Experience of Space and Place*, New York: St Martin's Press.

C40 Cities Climate Leadership Group, C40 Knowledge Hub (2020), How to build back better with a 15-minute city, https://www.c40knowledgehub.org/s/article/How-to-build-back-better-with-a-15-minute-city?language=en_US

CABE (2006), *The Value Handbook, Getting the Most from Your Buildings and Spaces*. London: Commission for Architecture and the Built Environment.

Campbell, T. (2012), *Beyond Smart Cities. How Cities Network, Learn, and Innovate*, London: Earthscan Publications.

Capolongo, S., Rebecchi, A., Dettori, M., Appolloni, L., Azara, A., Buffoli, M., Capasso, L., Casuccio, A., Oliveri, Conti G., D'Amico, A., Ferrante, M., Moscato, U., Oberti, I., Paglione, L., Restivo, V., D'Alessandro, D. (2018), Healthy design and urban planning strategies, actions, and policy to achieve Salutogenic cities, *International Journal of Environmental Research and Public Health*, 15(12): 2698.

Carmona, M. (2014), 'The place-shaping continuum, a theory of urban design process', *Journal of Urban Design*, 19 (1): 2–36.

Carmona, M. (2019a), 'Place value: Place quality and its impact on health, social, economic and environmental outcomes', *Journal of Urban Design*, 24(1): 1–48.

Carmona, M. (2019b), 'Principles for public space design, planning to do better', *Urban Design International*, 24(1): 47–59.
Carmona, M., Heath, T., Oc, T., Tiesdell, S. (2010), *Public Places-Urban Spaces*, Oxford: Architectural Press.
Carta, M. (2007), *Creative City*, Trento: List Lab.
Carta, M. (2014), *Reimagining Urbanism*, Trento: List Lab.
Carta, M. (2017), *Augmented City. A paradigm shift*, Trento: List Lab.
Castells, M. (1989), *The Informational City*, Oxford: Blackwell.
City of Portland (2015), *Climate Action Plan: Local Strategies to Address Climate Change*, Portland, OR: City of Portland.
City of Vancouver (2021a), *Green Vancouver*, https://vancouver.ca
City of Vancouver (2021b), *Parks, recreation and culture*, https://vancouver.ca
City of Vancouver (2021c), *VIVA Vancouver*, https://vancouver.ca
Ciuffi, V. (2013), 'Il Parco dei parchi: Superkilen di Copenhagen', *Abitare*, 516: 50.
Clout, H. (1984), 'Bordeaux: Urban renovation, conservation and rehabilitation', *Planning Outlook*, 27(2): 84–92.
Codarin, S. (2016), *Being in the City*, https://www.infobuildenergia.it
Comunian, R. (2009), Questioning creative work as driver of economic development: the case of Newcastle-Gateshead, *Creative Industries Journal*, 2(1): 57–71.
Corburn, J. (2009), *Toward the Healthy City. People, Places, and the Politics of Urban Planning*, Cambridge (MA): MIT Press.
CRA Associati (2018), *The Dynamic Street*, https://carloratti.com/project/the-dynamic-street/
CRA Associati (2019), *News from Paris: A Road to the Future*, https://www.thesmartcityjournal.com/en/cities/news-from-paris-a-road-to-the-future
Crappsley, R. (2017), 'Designing streets for good health: Now and in the future', *Urban Design Journal*, 143: 10–11.
Croucher, K., Myers, L. and Bretherton, J. (2007), *The Links between Greenspace and Health: A Critical Literature Review*, Stirling: Greenspace Scotland.
Cullen, G. (1961), *Townscape*, London: The Architectural Press.
Cushing, D. F., Miller, E. (2020), *Creating Great Places Evidence-Based Urban Design for Health and Wellbeing*, New York: Routledge.
D'Alessandro, D., Buffoli, M., Capasso, L., Fara, G. M., Rebecchi, A., Capolongo, S. (2015), 'Green areas and public health: Improving wellbeing and physical activity in the urban context', *Journal of Preventive Epidemiology*, 39 (4): 8–13.
Dannenberg, A. L., Frumkin, H., Jackson, R. J. (2011), *Making Healthy Places: Designing and Building for Health, Wellbeing and Sustainability*, Washington: Island Press.
David, H. (2006), *Sustainable New York City*, New York: New York City, Design Trust for Public Space and the New York City Office of Environmental Coordination.
Davies, J. S. (2002), 'The governance of urban regeneration: A critique of the "governing without government" thesis', *Public Administration*, *80*, 301–322.
Davoudi, S. et al. (2012), 'Resilience: A bridging concept or a dead end?', *Planning Theory and Practice*, 13 (2): 299–333.
de Magalhães, C., Carmona, M. (2009), 'Dimensions and models of contemporary public space management in England', *Journal of Environmental Planning and Management*, 52(1): 111–129.
De Roo, G., Porter, G. (eds.) (2007), *Fuzzy Planning – The Role of Actors in a Fuzzy Governance. Environment*, London: Ashgate, Aldershot and Routledge.

De Roo, G., Yamu, C., Zuidema, C. (2020), *Handbook on Planning and Complexity*, Cheltenham Glos UK: Edward Elgar Publishing.
de Vries, S., van Dillen, S. M. E., Groenewagen, P. and Spreeuwenberg, P. (2013), 'Streetscape greenery and health: Stress, social cohesion and physical activity as mediators', *Social Science and Medicine*, 94: 26–33.
de Vries, S., Verheij, R. A., Groenewegen, P. P., Spreeuwenberg, P. (2003), 'Natural environments – healthy environments? An exploratory analysis of the relationship between greenspace and health', *Environment and Planning A*, 35(10): 1717–1731.
de Waal, M. (2014), *The City as Interface. How Digital Media Are Changing the City*, Rotterdam: NAI010 Publishers.
Delgado, J. P. (2010), *Waterfront: The Illustrated Maritime History of Greater Vancouver*, Stanton Atkins, Vancouver: Dosil Publishers.
Depietri, Y., McPhearson, T. (2017), 'Integrating the grey, green, and blue in cities: Nature-based solutions for climate change adaptation and risk reduction', Kabisch, N., Korn, H., Stadler, J., Bonn, A. (eds.) *Nature-Based Solutions to Climate Change Adaptation in Urban Areas. Theory and Practice of Urban Sustainability Transitions*, Cham: Springer.
Desouza, K. C., Flanery, T. H. (2013), 'Designing, planning, and managing resilient cities: A conceptual framework', *Cities*, 35: 89–99.
Dines, N., Cattell, V., Gesler, W. and Curtis, S. (2006), *Public Spaces, Social Relations and Well-Being in East London*, Bristol: Policy Press.
Douglas, Margaret J. (2012), 'Making healthy places: Designing and building for health, wellbeing and sustainability', *Journal of Public Health*, 34 (2): 316.
Duany, A. Speck, J., Lydon, M. (2010), *The Smart Growth Manual*, New York: McGraw-Hill.
Dublin City Council (2003), *O'Connell Street: Special Planning Control Scheme*, Dublin: Dublin City Council.
Dublin Docklands Development Authority (2000), *Grand Canal Docks Planning Scheme, 2000*, Dublin: DDDA.
Dublin Docklands Development Authority (2006), *Grand Canal Docks Amending Planning Scheme 2006*, Dublin: DDDA.
EIU (2021), *The Global Liveability Index 2021*, https://www.eiu.com/topic/liveability
EIU-Economist Intelligence Unit (2021), *The Global Liveability Index 2021. How the Covid-19 pandemic affected liveability worldwide*, London: The Economist Intelligence Unit.
Evans, J., Jones, P. (2008), 'Rethinking sustainable urban regeneration: ambiguity, creativity, and the shared territory', *Environment and Planning A*, 40 (6): 1416–1434.
Fehr, R., Capolongo, S. (2016), 'Healing environment and urban health', *Epidemiol. Prev*. 40: 151–152.
Florida, R. (2002), *The Rise of the Creative Class: And How it's Transforming Work, Leisure, Community and Everyday Life*, New York: Perseus Book Group.
Florida, R. (2005), *Cities and the Creative Class*, New York: Routledge.
Forrester, J., Snell, C. (2007), 'Planning inclusive and sustainable urban regeneration: Balancing a visitor-based economy with local needs in the city of York, UK', *Journal of Urban Regeneration and Renewal*, 1(1): 69–82.
Friedman, A. (1997), 'Design for change: Flexible planning strategies for the 1990s and beyond', *Journal of Urban Design*, 2: 277–295.

Gallo, R. G., Townshend, T. G., Lake, A. A. (2015), 'Exlporing urban parks and their peripheral food environments using a case study approach: Young people and obesogenic environments', *Urban Design International*, 20(1): 28–43.

Garau, P., Lancerin, L., Sepe, M. (2015), *The Charter of Public Space*, Trento: LiST.

Gehl, J. (2010), *Cities for People*, Washington: Island Press.

Gehl, J. (2016), *Inclusive Healthy Places*, Copenhagen: Gehl Institute.

Gehl, J. (2020), *Public Space and Public Life During Covid 19*, Copenhagen: Gehl Institute.

Gospodini, A. (2002), 'European cities in competition and new 'uses' of urban design', *Journal of Urban Design* 7(1): 59–73.

Graham, S., Marvin, S. (2001), *Splintering Urbanism: Networked Infrastructures, Technological Mobilities and the Urban Condition*, London, New York: Routledge.

Grove, M., Zhang, T. (2020), 'Forged by floods: Wuhan Yangtze Riverfront Park', *Landscape Architecture Frontiers*, 8(3): 130.

Hafencity Hamburg GmbH (2004), *The Überseequartier: Pulsating centre on the Elbe River*, www.hafencity.com

Hafencity Hamburg GmbH (2006), *The Masterplan*, www.hafencity.com

Hafencity Hamburg GmbH (2009), *The birth of a city, Hafencity Hamburg GmbH.*

Hafencity Hamburg GmbH (2022), *A city of plazas, parks and promenades*, https://www.hafencity.com/en/urban-development/urban-spaces

Hall, P. (2014), *Good Cities, Better Lives: How Europe Discovered the Lost Art of Urbanism*, London: Routledge.

Hall, P., Preston, P. (1988), *The Carrier Wave: New Information Technology and the Geography of Innovation 1846–2003*, London: Unwin Hyman.

Hamburg Port Authority (2006), *Focus of dynamic growth markets*, Free and Hanseatic City of Hamburg, www.hamburg-port-authority.de

Hampton, K. N., Livio, O., Sessions, L. F. (2010), 'The social life of wireless urban spaces: Internet use, social networks, and the public realm', *Journal of Communication*, 60(4), 701–722.

Hatzelhoffer, L., Humboldt, K., Lobeck, M., Wiegandt, C. (eds.) (2012), *Smart City in Practice. Converting Innovative Ideas into Reality*, Berlin: JOVIS Verlag.

Healey, P. et al. (eds.) (1995), *Managing Cities: The New Urban Context*, Oxford: John Wiley.

Helliwell, John F., Richard, Layard, Jeffrey, Sachs, Jan-Emmanuel, De Neve (eds.) (2020), *World Happiness Report* 2020, New York: Sustainable Development Solutions Network.

Hernández, J. (2011), 'Accesibilidad y Seguridad', *Madrid Río. Un Proyecto de Transformación Urbana*, Madrid: Turner, 195–204.

Holling, C. (1973), Resilience and stability of ecological systems, *Annual Review of Ecology and Systematics*, 4: 1–23.

Jabareen, Y. (2013), 'Planning the resilient city: Concepts and strategies for coping with climate change and environmental risk', *Cities*, 31: 220–229.

Jabareen, Y. R. (2006), 'Sustainable urban forms: Their typologies, models, and concepts', *Journal of Planning Education and Research*, 26(1): 38–52.

Jacobs, J. (1961), *The Death and Life of Great American Cities: The failure of Modern Town Planning*, London: Peregrine Books.

Jones, P., Marshall, S., Boujenko, N. (2008), *Link and Place: A Guide to Street Planning and Design*, London: Landor Publishing.

Kaika, M., Swyngedouw, E. (2010), 'The urbanization of nature. Great promises, impasse, and new beginnings', in Bridge, G. E., Watson, S. (eds.) *New Companion to Urban Studies*, Oxford: Wiley/Blackwell.

Kaplan, R. and Kaplan, S. (1989), *Experience of Nature*, New York: Cambridge University Press.

Karvonen, A., Cugurullo, F., Caprotti, F. (eds.) (2019), *Inside Smart Cities: Place, Politics and Urban Innovation*, London, New York: Routledge.

Keith, N. H., Livio, O., Goulet, L. (2010), The social life of wireless urban spaces: Internet use, social networks, and the public realm, *Journal of Communication* 60: 701–722.

Kent, F. (2008), *Street as Place. Using Streets to Rebuild Communities.* New York: Project for Public Spaces.

Kyttä, M. et al. (2015), 'Urban happiness: Context-sensitive study of the social sustainability of urban settings', *Environment and Planning B*, 43(1): 34–57.

Law, A., Qin, Q. (2013), Searching for economic and cosmopolitan roots: The historical discourse of "Hankou Merchant Port Nostalgia" in The central Chinese city of Wuhan, *METU JFA*, 35(2): 173–195.

Lawrence, R. J. (2008), 'Urban environmental health indicators: Appraisal and policy directives', *Rev Environ Health*, 23:299–325.

Lawton, P., E. Murphy, D. Redmond (2010), 'The role of 'creative class' ideas in urban and economic policy formation: the case of Dublin, Ireland', *International Journal of Knowledge-Based Development*, 1(4), 267–286.

Lehmann, S. (2010), 'Green urbanism: Formulating a series of holistic principles', *SAPI EN. S. Surveys and Perspectives Integrating Environment and Society*, 3 (2), http://journals.openedition.org/sapiens/1057

Lewis Sam Wren (2016), *Happy City Index 2016*, https://www.centreforthriving-places.org/wp-content/uploads/2016/10/Happy-City-Index-2016-Report-FINAL.pdf

Liu, Q. (2012), The practice and exploration of dynamic conservation historic heritage planning in Wuhan city. Liu, Y., Lavroff, S (eds.) *Hankou French Concession and its Architecture*, Wuhan: Wuhan Press.

Liu, Q. Z., Wu, Z. L., He, M., Huang, N. (2004), Landscape construction in urban riverside area: Planning and design of the Hankou riverside project in Wuhan City, *City Planning Review*, 28(3): 83–87.

London, F. (2020), *Healthy Placemaking Wellbeing Through Urban Design*, London: Routledge.

Lowe M, Whitzman C, Badland H, et al. (2015), 'Planning healthy, liveable and sustainable cities: How can indicators inform policy?' *Urban Policy Res*, 33: 131–144.

Lydie, P. (2003), 'Bordeaux les deux rives: Quelle(s) perspective(s)' in Patrick, B, Thierry, P., *L'Urbain et ses imaginaires*, Pessac: Maison des Science de l'Homme Aquitaine.

Lynch, K. (1960), *The Image of the City*, Boston: MIT Press.

Lynch, K. (1984), *Good City Form*, Boston: MIT Press.

Maas, J., Verheij, R. A., Groenewegen, P. P., de Vries, S. and Spreeuwenberg, P. (2006), 'Green space, urbanity, and health: How strong is the relation?', *Journal of Epidemiology and Community Health*, 60(7): 587–592.

Madanipour, A (ed.) (2011), *Whose Public Space?: International Case Studies in Urban Design and Development*, London: Routledge.

Madanipour, A., Knierbein, S., Degros, A. (2014), *Public Space and the Challenges of Urban Transformation in Europe*, New York: Routledge.

Mareggi, M. (2020), *Spazi Aperti. Ragioni, progetti e piani urbanistici*, Milano: Planum Publisher.

Martone, A., Sepe, M. (2012), 'Creativity, urban regeneration and sustainability: The Bordeaux case study', *Journal of Urban Regeneration and Renewal*, 5(2), 164 -183.

Mateo, S. et al. (2012), *Nice Côte d'Azur labellisée ÉcoCité*, www.ecovallee-plaineduvar.fr

Mathewson, D., Karaga, K. (2018), 'Vancouver as a benchmark', *Urban Design Journal*, 145: 25–27.

McCay, L. (2017), 'Designing mental health into cities', *Urban Design Journal*, 142: 25–27.

McHarg, I. (1969), *Design with Nature*, New York: Doubleday.

McKenzie, M., Hutton, T. (2015), 'Culture-led regeneration in the post-industrial built environment: Complements and contradictions in victory square, Vancouver', *Journal of Urban Design*, 20 (1): 8–27.

Meng, J., Liu, X. (2007), Hankou river beach flood control and environmental comprehensive harnessing project, *Urban Roads Bridges & Flood Control*, (6): 61–65.

Mercer (2019), '*Quality of living city ranking*', https://mobilityexchange.mercer.com/insights/quality-of-living-rankings

Mitchell, R., Popham, F. (2008), 'Effect of exposure to natural environments on health inequalities: An observational population study', *The Lancet*, 372: 1655–1660.

Mitchell, W. J. (1995), *City of Bits: Space, Place, and the Infobahn*, Cambridge, MA: MIT Press.

Montgomery, C. (2013), *Happy City, Transforming Our Lives through Urban Design*, London: Penguin Books.

Montgomery, J. (1995),' The story of temple bar: Creating Dublin's cultural quarter', *Planning Practice and Research* 10 (2): 135– 172.

Montgomery, J. (1998), 'Making a city: Urbanity, vitality and urban design', *Journal of Urban Design*, 3: 93–116.

Moore-Cherry, N., Vinci, I. (2012), 'Urban regeneration and economic crisis: Past development and future challenges in Dublin, Ireland, Planum', *The Journal of Urbanism*, 25(2): 1–16.

Moreno, C. (2020), *Et après ? #30 Vie urbaine et proximité à l'heure du COVID-19*, Paris: Edition de L'Observatoire.

Musterd, S., Kovács, Z. (2013), *Place-making and Policies for Competitive Cities*, Oxford: Wiley-Blackwell.

Neal, Z. (2012), *The Connected City*, New York, NY: Routledge.

Newcastle and Gateshead Council (2015), *Gateshead Quays scheme*, https://www.gateshead.gov.uk/article/16918/Planning-approval-for-260m-Gateshead-Quays-scheme

Nijkamp, P., Perrels, A. H. (1994), *Sustainable Cities in Europe*, London: Earthscan.

Office of the Deputy Prime Minister (2004), *Planning Policy Statement 1: Delivering Sustainable Development*, London: ODPM.

Ojerio, R., Moseley, C., Lynn, K., Bania, N. (2011), 'Limited involvement of socially vulnerable populations in federal programs to mitigate wildfire risk in Arizona', *Natural Hazards Review*, 12 (2): 28–36.

Otgaar, A., Klijs, J., van den Berg, L. (2011), *Towards Healthy Cities Comparing Conditions For Change*, Ashgate: Routledge.

Pendlebury, J., Porfyriou, H. (2017), 'Heritage, urban regeneration and place-making', *Journal of Urban Design*, 22(4): 429–432.

Pickett, S. T. A., Cadenasso, M. L., Grove, J. M. (2004), 'Resilient cities: Meaning, models, and metaphor for integrating the ecological, socio-economic, and planning realms', *Landscape and Urban Planning*, 69: 369–384.

Pineo, H., Glonti, K., Rutter, H. et al. (2018), 'Urban health indicator tools of the physical environment: A systematic review', *J Urban Health* 95: 613–646.

Pizzo, B. (2015), 'Problematizing resilience: Implications for planning theory and practice', *Cities*, 43: 133–140.

Porfyriou, H., Sepe, M. (eds.) (2017), *Waterfront Revisited*, London, New York: Routledge.

Potesta, G. (2019), 'Smart cities and place making: The "Sense of Place" in the implementation of smart cities in the Arabian Gulf', in Samad, W., Azar, E. (eds.) *Smart Cities in the Gulf*, Singapore: Palgrave Macmillan.

Punter, J. (2003), *The Vancouver Achievement: Urban Planning and Design*, Vancouver: UBC Press.

Pykya, R. et al. (2018), 'Individual and environmental factors associated with green exercise in urban and suburban areas', *Health and Place*, 55: 20–28.

Ratti, C. (2014), *Architettura Open Source*, Einaudi: Torino.

Rockefeller Foundation, Arup (2019), *City Resilience Index. Understanding and Measuring City Resilience*, https://www.cityresilienceindex.org

Ruijsbroek, A., Wong, A., Kunst, A. E., van den Brink, C., van Oers, H. A. M., Droomers, M., et al. (2017), 'The impact of urban regeneration programmes on health and health-related behaviour: Evaluation of the Dutch district approach 6.5 years from the start', *PLoS One*, 12(5): 1–8.

Russo, M. (2018), 'Rethinking resilience, design the city through its metabolism', *TECHNE, Journal of Technology for Architecture and Environment*, 15: 39–44.

Sarkar, C., Webster, C., Gallacher, J. E. (2014), *Healthy Cities: Public Health through Urban Planning*, Cheltenham: Edward Elgar Publishing.

Sassen, S., Kourtit, K. (2021), 'A post-corona perspective for smart cities: 'Should I Stay or Should I Go?'' *Sustainability*, 13: 9988.

Saunders, L. (2017), 'Healthy cities', *Urban Design Journal*, 142: 12–39.

Sepe, M. (2022), 'Public spaces and the COVID-19 pandemic: two best practices in the Mediterranean area', in S. Capasso, G. Canitano (Eds.), *Mediterranean Economies 2022*, Milano: il Mulino.

Sepe, M. (2021), 'Covid-19 pandemic and public spaces: Improving quality and flexibility for healthier places', *Urban Design International* 26(2): 159–173.

Sepe, M. (2020a), 'Regenerating places sustainably: the healthy urban design', *International Journal of sustainable development and planning*, 15(1): 14–27.

Sepe, M. (2020b), *Spazi pubblici nella città contemporanea*, Roma: Inu Edizioni.

Sepe, M. (2019a), '*Liveable and Healthy City Design*', *Wit Transactions on Ecology and the Environment*, Southampton: Wit Press.

Sepe, M. (2019b), 'Measuring the impact of slow regeneration process: The HLH Pl@ce index', *UPLAND*, 2 (4): 75–84.

Sepe, M. (2018), 'Place identity and creative district regeneration: the case of 798 in Beijing and M50 in Shanghai Art zones', *Metu Journal of the Faculty of Architecture*, 35(2): 151–171.

Sepe, M. (2017a), 'The role of public space to achieve urban happiness', *International Journal of Sustainable Development and Planning*, 2(4): 724–733.

Sepe, M. (2017b), 'Placemaking, liveability and public spaces: Achieving sustainability through happy places', *Journal of Public Space*, 11(2): 63–76.

Sepe, M. (2016), 'Industrial regeneration and cultural identity: The case of 798 Art Zone and M50 Art District', *Urbanistica*, 158: 106–114.

Sepe, M. et al. (2016), *Italy's Presidency of the Council of Ministers Habitat III Italy's National Report*, http://www.governo.it/sites/governo.it/files/UN_HABITAT_III_ITALY_NATIONAL_REPORT_EN.pdf

Sepe, M. (2015a), 'Reti di luoghi, paesaggi delle tecnologie e nuove connessioni', *Crios*, 10(2): 69–80.

Sepe, M. (2015b), 'Improving sustainable enhancement of cultural heritage: Smart placemaking for experiential paths in Pompeii', *International Journal of Sustainable Development and Planning*, 10(5): 713–733.

Sepe, M. (2014a), 'Urban transformation,socio-economic regeneration and participation: Two case of creative urban regeneration', *International Journal of Urban Sustainable Development*, 6(1): 20–41.

Sepe, M. (2014b), 'Preserving, reconstructing and enhancing the place identity: Identifying principles of urban sustainability with PlaceMaker method', *International Journal of Sustainable Development and Planning*, 9(2): 277–300.

Sepe, M. (2013a), *Planning and Place in the City: Mapping Place identity*, London-New York: Routledge.

Sepe, M. (2013b), 'Urban history and cultural resources in urban regeneration: A case of creative waterfront renewal', *Planning Perspectives Journal*, 28(4): 595–613.

Sepe, M., Pitt, M. (2013), 'Improving liveability and attractiveness by preserving place identity in emblematic thoroughfares: A method and a case study', *Urban Design International*, 18 (3): 229–249.

Sepe, M. (2012), 'A dynamic method and interactive software to monitor and design place identity', in G. Borruso, S. Bertazzon, A. Favretto, B. Murgante, C. M. Torre (eds.) *Geographic Information Analysis for Sustainable Development and Economic Planning: New Technologies*, Hershey, PA: IGI Global.

Sepe, M. (2010), 'Place identity and PlaceMaker: planning the urban sustainability', special issue "Best Practices on Land Management Strategies", *Journal of Urban Planning and Development*, 136 (2): 139–146.

Sepe, M. (2009a), 'Creative urban regeneration between innovation, identity and sustainability, special issue L. Fusco Girard, P. Lombardi, P. Nijkamp (eds.) 'Creative Urban Design and Development'', *International Journal of Sustainable Development*', 12(2–3–4): 144–159.

Sepe, M. (2009b), 'PlaceMaker method: Planning walkability by mapping place identity', *Journal of Urban Design*, 14(4): 463–487.

Shu, Y., Qin, J. (2014), 'Resilient development of urban waterfront in fast-changing cities—A case study of Wuhan, China', *Journal of Civil Engineering and Architecture*, 8(8): 1073–1083.

Sim, D. (2019), *Soft City: Building Density for Everyday Life*, Washington, DC: Island Press.

Skelton, C. (2014), *Soft City Culture and Technology: The Betaville Project*, New York, Heidelberg, London: Springer.

Souter-Brown, G. (2014), *Landscape and Urban Design for Health and Well-Being: Using Healing, Sensory and Therapeutic Gardens*, London, New York: Routledge.

Speck, J. (2012), *Walkable City, How Downtown Can save America, One Step at a Time*, New York: North Point Press.

Sudjic, D. (2016), *The Language of Cities*, Milton Keynes, UK: Penguin.

Talukder, S, Capon, A, Nath, D, Kolb, A, Jahan, S, Boufford, J. (2015), 'Urban health in the post-2015 agenda', *Lancet*, 385(9970): 769.
Taylor, A. F., Kuo, F. E., Sullivan, W. C. (1998), 'Growing up in the inner city: Green spaces as places to grow,' *Environmental Behavior*, 30(1): 3–27.
Thompson, C., Aspinall, P. A. (2011), 'Natural environments and their impact on activity, health, and quality of life', *Applied Psychology: Health and Well-Being*, 3(3): 230–260.
Tsiomis, Y., Ziegler, V. (2007), *Anatomie de projets urbains, Bordeaux, Lyon, Rennes, Strasbourg*, Paris: Editions de la Villette.
Towers, G. (2005), *Introduction to Urban Housing Design: At Home in the City*. Oxford: Architectural Press.
Townsend, A. M. (2013), *SMART CITIES: Big Data, Civic Hackers, and the Quest for a New Utopia*. New York: W. W. Norton & Company.
Ulrich, R., Simons, R., Losito, B., Fiorito, E., Miles, M., Zelson, M. (1991), 'Stress recovery during exposure to natural and urban environments', *Journal of Environmental Psychology*, 11(2): 201–230.
UN Department of Economic and Social Affairs – Sustainable Development (2015), 17 Sustainable Development Goals, https://sdgs.un.org/goals
UN-Habitat (2013), *Global Public Space Toolkit From Global Principles to Local Policies and Practice*. Nairobi: United Nations Human Settlements Programme.
UN-Habitat (2016), *The New Urban Agenda*, https://habitat3.org/the-new-urban-agenda
Vale, J. L. (2014), 'The politics of resilience cities: Whose resilience and whose city?', *Building Research and Information*, 42 (2): 191–201.
Vale, L. J., Campanella, T. J. (2005), *The Resilient City: How Modern Cities Recover from Disaster*, Oxford Scholarship Online.
Victoria State Government (2017), *20 Minute Neighbourhoods*, https://www.planning.vic.gov.au/policy-and-strategy/planning-for-melbourne/plan-melbourne/20-minute-neighbourhoods
Ward Thompson, C. (2011), 'Linking landscape and health: The recurring theme', *Landscape and Urban Planning*, 99(3–4): 187–195.
Weijers, E. P., de Vries, B., Bosch, P., van Dinther, D., van der Meulen, S., Gehrels, H., Schasfoort, F., Broslma, R., Geerling, G., Goossens, M., Jacobs, C., de Jong, M., Kok, S., Massop, H., Osté, L., Pérez-Soba, M., Rovers, V., Smit, A., Verweij, P. (2016), *Designing Green and Blue Infrastructure to Support Healthy Urban Living*, Petten: ECN.
Whyte, W. H. (1980), *The social Life of small public space*, Washington DC: Conservation Foundation.
Wilson, E. O. (1984), *Biophilia, the Human Bond with Other Species*, Cambridge, MA: Harvard University Press.
Wong, C. (2006), *Indicators for Urban and Regional Planning: The Interplay of Policy and Methods*, Abingdon: Routledge.
World Health Organization (2014), *The Urban Health Index: A Handbook for its Calculation and Use*, Kobe: WHOS.
Xu, H. (2011), 'The holistic urban planning approach of urban sustainable development', *Advanced Materials Research*, 280: 58–61.
Zhai, Y., Li, K., Liu, J. (2018), 'A conceptual guideline to age-friendly outdoor space development in China: How do Chinese seniors use the urban comprehensive park? A focus on time, place, and activities', *Sustainability*, 10: 3678.

Zheng, H. W., Shen, G. Q., Song, Y., Sun, B. X., Hong, J. K. (2017), 'Neighbourhood sustainability in urban renewal: An assessment framework', *Environment and Planning B-Planning and Design*, 44(5): 903–924.

Zlot, A., Schmid, T. (2005), 'Relationships among community characteristics and walking and bicycling for transportation or recreation', *American Journal of Health Promotion*, 19(4): 314–317.

Zolli, A., Healey, A. M. (2012), *Resilience. Why Things Bounce Back*, New York: The Free Press.

## Websites

http://www.hafencity.com

http://www.hamburg-port-authority.de

http://www.waterfront-net.org/europa/index.htm http://www.waterfront communitiesproject.org

https://www.b-m-a.fr

https://www.c-mw.net/on-the-newcastlegateshead-waterfront-big-reveal

https://www.hafencity.com/quartiere/am-lohsepark

https://www.madrid.es/portales/munimadrid/es/Inicio/El-Ayuntamiento/Arganzuela/Direcciones-y-telefonos/Parque-Madrid-Rio/

https://www.meetnewcastlegateshead.com/newcastlegatesheadquays

https://www.ngquays.com/landscaping

www.nice.fr

www.nicecotedazur.org

www.realdania.dk

# Index

Pages in *italics* refers figures and **bold** refers tables.